Electrics Afloat

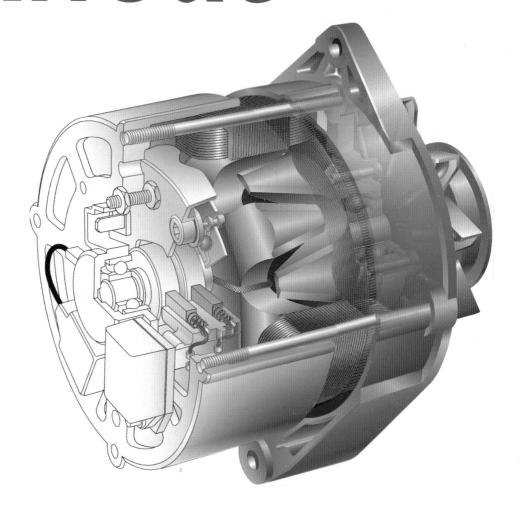

Alastair Garrod

Published 2002 by
Adlard Coles Nautical
an imprint of A & C Black Ltd,
37 Soho Square, London, WID 3QZ
www.adlardcoles.co.uk

ISBN 0-7136-6149-6

First edition published 2002
Reprinted 2005

Typeset in Stone Sans 12pt & 8pt
Printed and bound in Italy by
G. Canale & Co. S.p.a.

Contents

The Basics

Batteries

Start Your Engines

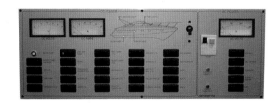

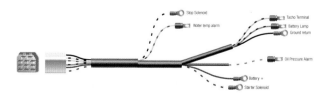

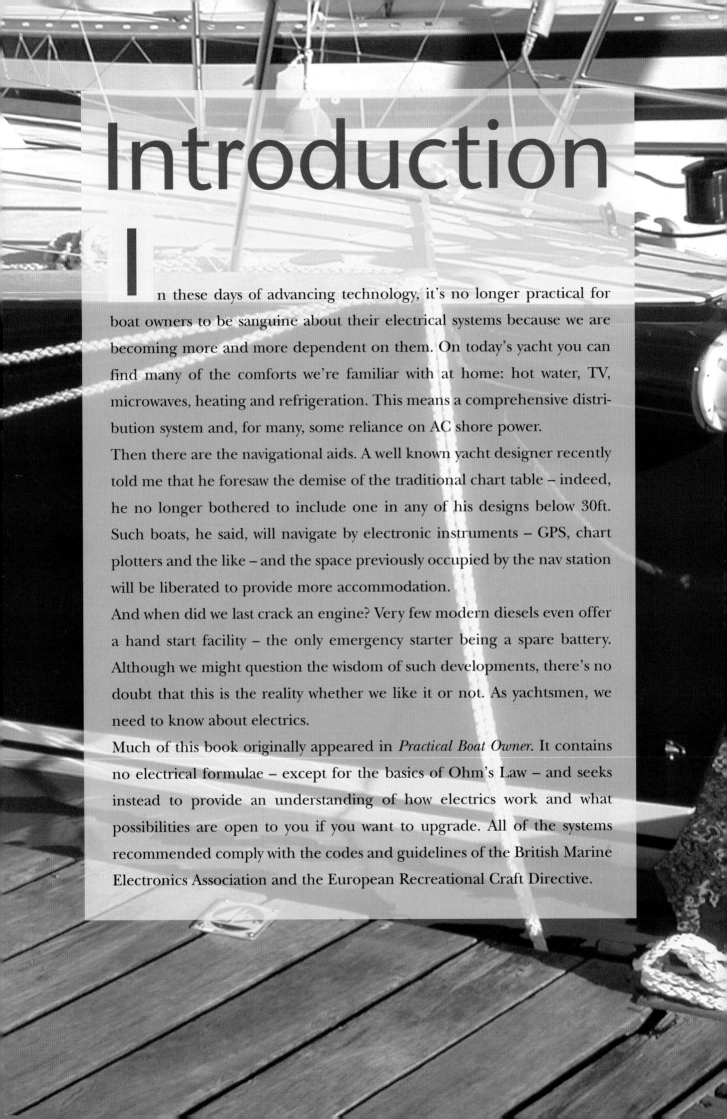

Introduction

In these days of advancing technology, it's no longer practical for boat owners to be sanguine about their electrical systems because we are becoming more and more dependent on them. On today's yacht you can find many of the comforts we're familiar with at home: hot water, TV, microwaves, heating and refrigeration. This means a comprehensive distribution system and, for many, some reliance on AC shore power.

Then there are the navigational aids. A well known yacht designer recently told me that he foresaw the demise of the traditional chart table – indeed, he no longer bothered to include one in any of his designs below 30ft. Such boats, he said, will navigate by electronic instruments – GPS, chart plotters and the like – and the space previously occupied by the nav station will be liberated to provide more accommodation.

And when did we last crack an engine? Very few modern diesels even offer a hand start facility – the only emergency starter being a spare battery. Although we might question the wisdom of such developments, there's no doubt that this is the reality whether we like it or not. As yachtsmen, we need to know about electrics.

Much of this book originally appeared in *Practical Boat Owner*. It contains no electrical formulae – except for the basics of Ohm's Law – and seeks instead to provide an understanding of how electrics work and what possibilities are open to you if you want to upgrade. All of the systems recommended comply with the codes and guidelines of the British Marine Electronics Association and the European Recreational Craft Directive.

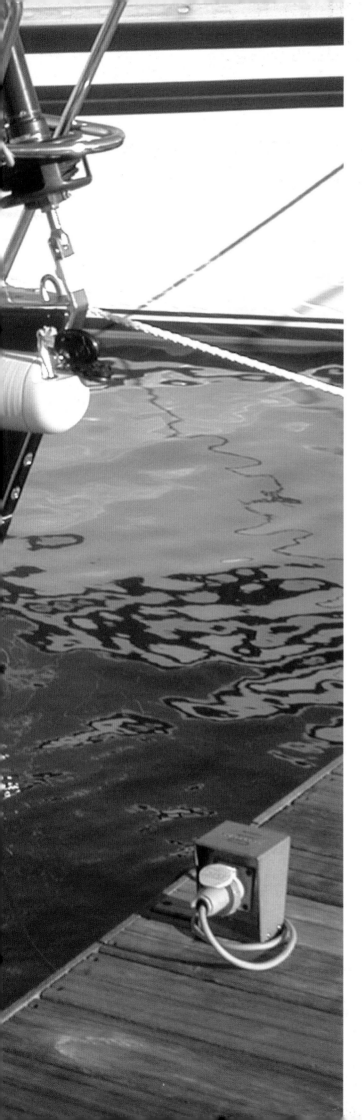

Acknowledgements

When I started researching for this book, I very quickly realised that if you filled a room with twenty electricians and put a simple question to each of them, you'd probably leave the room twenty times more confused than when you went in. It therefore became obvious that I needed to select my sources and consultants very carefully. In my search, the same names kept emerging and I wish to credit these few with my heartfelt thanks for their outstanding technical support and patience.

Jack Monk and Danny Jones, then of Mastervolt UK. Danny Jones sits on the BMEA.

Michael Taplin, founder of Taplin International and now retired.

John St Bickley of ETA circuit breakers. John also sits on the BMEA.

Steve Barnes of S.D.Barnes Associates, Poole.

Tony Johns, Secretary to the BMEA who also features regularly in *Practical Boat Owner*.

George Huxtable, who applied much of his knowledge and experience in technically vetting and proofing the original 'Electrics Made Simple' series in *Practical Boat Owner*.

Lastly, I would like to thank my colleague and Associate Editor at *PBO*, Andrew Simpson, who's been a rock throughout and painstakingly sub-edited his way through my scrawls and ultimately managed to apply some rhythm to my words. Also much of his technical input has been featured in this book – many thanks Andrew.

A considerable element of the Marine Electrical trade supplied me with their product photographs and technical drawings. Those companies include: Mastervolt UK, Merlin Equipment, Aqua Marine, Plastimo, Volvo UK, Delphi Auto Batteries, Ampair, Bosch and Vetus.

Their co-operation is greatly appreciated.

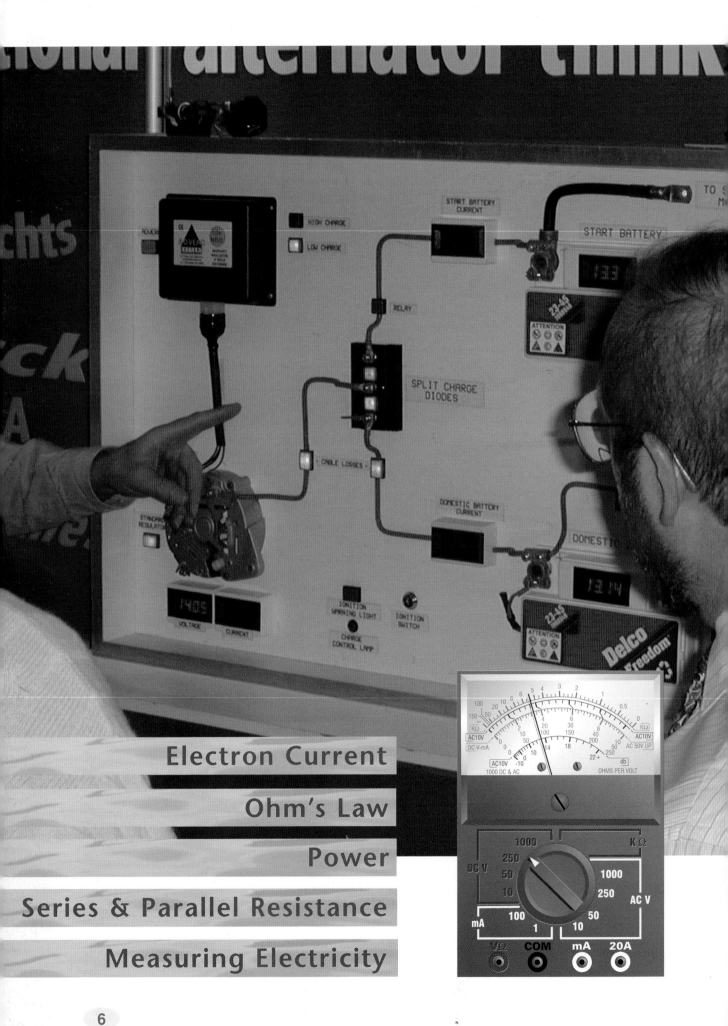

Electron Current

Ohm's Law

Power

Series & Parallel Resistance

Measuring Electricity

The Basics

Boat electrics become very much less confusing once the fundamentals are understood. Unfortunately, many boat owners find themselves with their schooldays well behind them, and much of what they learned just a distant or distorted memory. So, for those who need reminding or have never covered the subject before, this first chapter will take you through the basics without complicated mathematics or theory. Once you understand the principles, the rest will be easy to assimilate.

Current

Electricity comes from the movement of electrons within the atomic structure of a conductive material (see below). The electrons move from one atom to another, creating the motion we know as electrical current.

It's easiest to think of this as being like a fluid flowing along the conductor – the flow rate of which is called the electrical CURRENT and is measured in *amperes* (often abbreviated as amps or A). Smaller currents are measured in milliamps or thousandths of an amp.

Voltage

But the current can't flow until acted upon or initiated by an electrical force or pressure. This pressure is called VOLTAGE and is measured in *volts* (V).

Resistance

Any obstruction to the flow of current is called RESISTANCE and is measured in *ohms* (O).

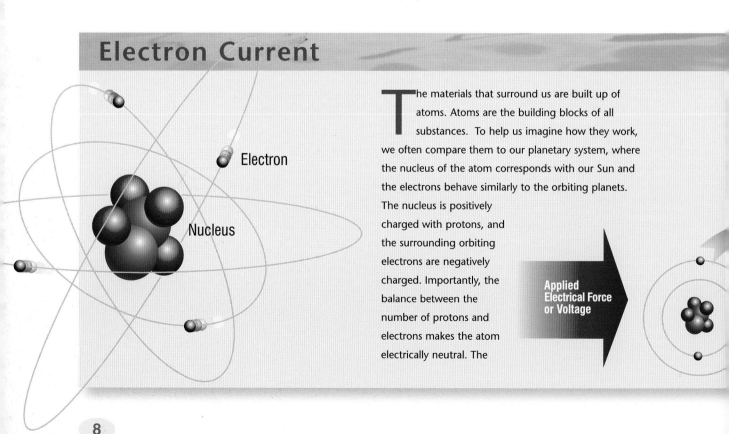

Electron Current

Electron

Nucleus

The materials that surround us are built up of atoms. Atoms are the building blocks of all substances. To help us imagine how they work, we often compare them to our planetary system, where the nucleus of the atom corresponds with our Sun and the electrons behave similarly to the orbiting planets. The nucleus is positively charged with protons, and the surrounding orbiting electrons are negatively charged. Importantly, the balance between the number of protons and electrons makes the atom electrically neutral. The

Applied
Electrical Force
or Voltage

Ohm's Law

The three elements of voltage, current and resistance, are closely interrelated and in choosing any two of the three values we will see how this relationship develops. Firstly, for a given resistance, any increase in the electrical pressure or voltage would increase the electrical flow of current by a proportional amount. Secondly, for a given voltage, any increase in the resistance to the path of any current would decrease or impede the electrical flow of current by a proportional amount.

These relationships between the combination of voltage, current and resistance form the first basic fundamental law of electricity. That is called Ohm's Law which can be simply shown as

$$VOLTAGE = CURRENT \times RESISTANCE$$
$$or \ VOLTS = AMPS \times OHMS$$

Given any two values, then the third can easily be calculated.

A heavy duty terminal post. The post itself will be of a good conductive material, like steel or brass. The insulation cap will be of rubber or plastic, which is a non-conductive material.

characteristic that distinguishes a conductive from a non-conductive material is the fact that the electrons in the conductive material can break away to join the orbits of other atoms. However, it takes an electrical force or VOLTAGE to dislodge an electron from its atom – and, in doing so, upsets its electrical neutrality. This leaves the atom with a net positive charge, thus attracting a free negatively charged electron to take its place. The movement of electrons hopping from one atom to another represents the motion of electrical current. We can imagine this movement as being like a pea shooter filled with a long line of peas – as a pea (an electron) is inserted at one end, the whole line moves along one to dislodge a pea at the other end, which then goes back to the beginning of the queue, and so on.

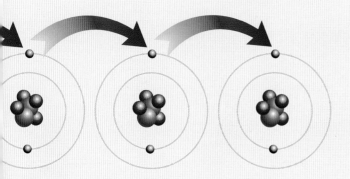

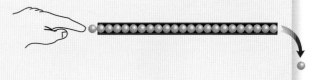

Power

The fourth element is power, which can be defined as the rate in which energy is delivered or consumed and is measured in watts (W). It follows therefore that if the amps (A) represent the rate at which the current is delivered and volts (V) represent the force or electrical pressure of delivering this current, then the power can be defined as the product of both the voltage and current put together.

Power and Resistance

By applying the formula above, we can make a very valuable observation which has considerable relevance to boat electrics.

A 20 watt light bulb burning in a 240V domestic household system will consume 83 milliamps of current. But if a 20 watt light bulb was burning in a boat's 12V domestic system it would consume 1666 milliamps of current.

So, for the same powered light bulb, the boat's electrics will be circulating a current 20 times that of a house – to power a device of any given wattage.

This highlights a potential danger. Most of us are very wary of 240V, knowing this can electrocute us, and are fairly relaxed about 12V, knowing that this can't. But what's not so widely appreciated is the fire risk that low voltage systems bring. Higher currents mean that the electrons hopping from atom to atom are more agitated and moving faster within the conductor. And in doing so they will suffer effects which can be thought of as

internal friction, thereby producing a considerable amount of heat. This analogy of internal friction is, of course, nothing more than resistance, perhaps in the form of unwelcome resistance in electrical feed wires. The effects will be further exaggerated if an intentional resistance, like an electrical load, is introduced or the wire has too small a cross-sectional area and the movement of the atoms is further confined. In this last case, power will be lost overcoming the friction, and as it can't go anywhere the spent power will be released in the form of heat.

This phenomenon is used beneficially in everyday life, and can be seen in such things as electric bar heaters, cookers, kettles and even the simple light bulb. Under controlled conditions we deliberately pass a current through a high resistance wire, so that the internal friction of the electrons will give us heat – in the case of a light bulb, enough heat for the fine filament wire to glow white hot.

But what might be a good thing for

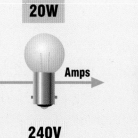
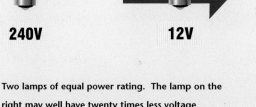

20W **240V** Amps

20W **12V** x 20 Amps

Two lamps of equal power rating. The lamp on the right may well have twenty times less voltage across it than the lamp on the left, but it's going to be carrying twenty times the current.

Therefore:

POWER (watts) = VOLTAGE x CURRENT

or   WATTS = VOLTS x AMPS

These fundamental relationships between voltage, current, resistance and power are central to any understanding of electrical systems.

heaters and light bulbs isn't at all what we want with our electrical conductors, where unwanted internal resistance could cause the wire to get hot, possibly melting the insulation and even causing a fire. For marine 12V systems this means that we must carefully match the sizes of all cables to the currents we expect them to carry. We must also understand that undersized cables will offer greater resistance to current flow, and will result in a voltage drop along the length of the cable. The practical effects of this would be that any equipment downstream won't receive the full battery voltage.

The engine battery voltmeter plummets as 400 amps of starting current rushes through large cables and the low resistance offered by the starter motor.

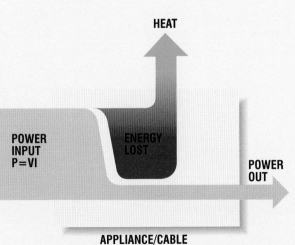

If an appliance or a cable possesses any form of resistance to current, then energy will be lost to overcome this resistance. This energy will be released as heat – perhaps enough to melt the insulation of electrical feed wires.

When we have conductor cables of ample cross-sectional area and negligible resistance supplying a load of equally low resistance, it's possible for the current flow to be so rapid that the battery can't keep up. A good example is when you start your engine. The cables that feed the starter motor are very large and the motor itself offers such little resistance that it will exceed the battery's ability to maintain the system voltage. Keep an eye on the voltmeter when you turn the key and you'll see the voltage fall from twelve to about ten and a half volts – perhaps even lower.

Simple Circuits

Here we have a battery supplying electrical current to a load – a light bulb in this case. The circuit requires a pair of cables, positive and negative. Convention suggests that we think of the direction of current flow as coming out from the positive battery terminal, through the circuit and then back to the negative battery terminal.

This lamp will of course have a certain amount of resistance. Resistances in an electrical circuit can be placed either in series or in parallel, or as a combination of both. The arrangement of resistances chosen will have profound effects on the voltages and currents within the circuit.

Series Resistance

A string of lamps or other sources of resistance in series will offer progressive impediments to the current flow in any circuit. According to their resistance, there will be a drop in the electrical pressure or voltage at every lamp, and the sum of all these voltage drops will equate to the battery voltage. The current through each lamp is the same – indeed it's the same at any point in the circuit and the amount of current depends on the total resistance of all the lamps. This means that for every lamp that's added in series (see below), the overall current (I TOTAL) in the circuit will be reduced, thus dimming the glow of all the lamps. This is rather like tying a series of loose knots along the length of a garden hose. With every knot you tie, the water flow will become a little more constricted until there's nothing but a trickle coming out of the end.

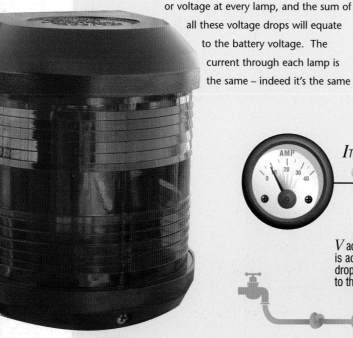

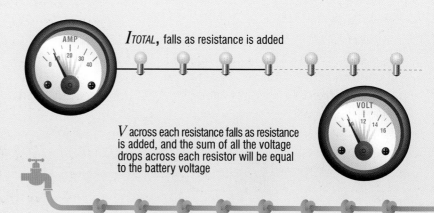

ITOTAL, falls as resistance is added

V across each resistance falls as resistance is added, and the sum of all the voltage drops across each resistor will be equal to the battery voltage

Ball cock

Loft tank

Parallel Resistance

Although series resistance is straightforward and logical, the results of putting our lamps in parallel aren't quite so obvious. For, the more lamps you add in parallel, the more the overall current (I TOTAL) increases while the glow of each lamp remains unchanged.

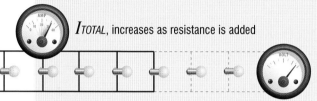

ITOTAL, increases as resistance is added

V across each resistance remains the same

At first glance this would seem to defy logic. How can it be that adding resistors in parallel will lower the circuit's resistance and also increase the load?

To explain this, let's leave electricity and look at a typical household domestic water system where we have a large water tank in the loft which stores and maintains a head of water – similar in some ways to a battery.

Now let's suppose we open a sink tap in the bathroom a turn and allow the water to drain away – this could be likened to switching on a lamp. The water level in the tank starts to fall slowly, enough to open the tank ball cock slightly and allow the water in the rising main to replenish it. Soon, a state of balance will exist.

Then run another tap. The tank's water level drops further and the ball cock opens wider; by now you may even be able to hear it from downstairs. Run a bath as well and the rising main will be heard to roar since the ball cock will be fully open as it struggles to maintain the level in the tank.

What we're doing in effect is reducing the resistance to the water flowing from the tank by providing more and more paths along which it can travel and in doing so, making the rising main work harder – if the head of water is to be maintained. At some point, through opening more taps, demand will exceed supply and the mains won't be able to replenish the head of water fast enough. At this point we could say that the system has become overloaded – a similar situation to the one in our starter motor circuit (see P. 10).

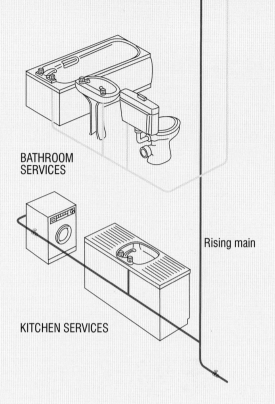

BATHROOM SERVICES

KITCHEN SERVICES

Rising main

Household plumbing can generally be described as a network of parallel circuits and the events that have just been described resemble very closely what happens when resistors are added in parallel in an electrical circuit.

The action of resistors in an electrical circuit can be tested by experimentation by first placing a pair of identical resistors in series across a voltage and measuring the current, then connecting them in parallel across the same voltage and measuring the current again. You will find that the second reading shows four times the current as the first.

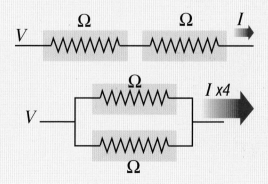

Parallel Resistance

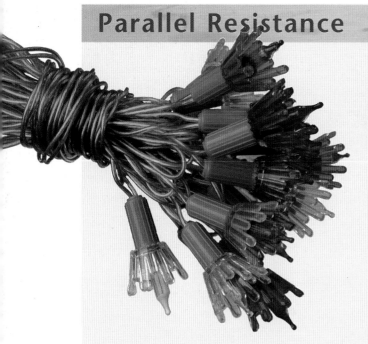

That's to say, each branch is directly connected to the battery and each branch will receive an equal voltage from the battery. The currents from each of the independent branches all join and add together to provide a sum of total current.

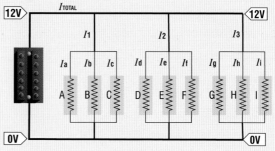

$I1$ is the sum of Ia, Ib and Ic and the same principles apply to $I2$ and $I3$
I TOTAL is the sum of $I1$, $I2$ and $I3$
The voltage across each appliance is common throughout.

Nearly all the electrical circuits on a boat will be in parallel to the supply, and there are a number of reasons why this should be so. Christmas lights provide a good example of why we need parallel circuits. These usually consist of twenty-four 10V lamps, connected in series and powered from

Circuit interrupted

a 240V supply. As many of us have discovered, when one lamp fails the whole chain is extinguished. But if, instead, we had connected twenty-four 240V lamps in parallel, the failure of one (or more) would no longer

Circuit uninterrupted elsewhere

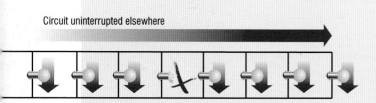

affect the rest. Obviously, for a boat's electrical system, this is the better and safer practice. For when there are several circuits in parallel, each one is independent of the others, and although the currents in the various branches may be different, the potential difference or voltage between the common ends must be the same.

The general exception to the practice of paralleling is when circuit breakers (CBs), local switches and fuses are added to the system.
Local switches placed in series to each of the parallel devices offer localised on/off switching without affecting other systems or circuits.
CBs or fuses, on the other hand, are usually placed in series to the paralleled network – their purpose, should it be necessary, being to remove power from the whole network.

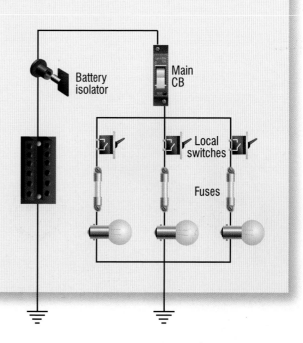

Battery isolator

Main CB

Local switches

Fuses

Measuring Electricity

There's a wide selection of electrical meters on the market at prices to suit every pocket. The first choice you should make is whether you want an analogue or digital display readout – largely a matter of personal preference, though digitals are rather more rugged.

Fixed or fitted displays on distribution panels are dealt with later, but every boat should carry a portable meter, capable of measuring amps, volts and ohms (AVO for short). Such an instrument will act as your stethoscope, and not only will it help you trace any faults, but it will tell you if circuits are safe to work on as well.

A good AVO meter for a boat should be capable of measuring 0 to 500V AC/DC, 0 to 10A and 0 to infinity Ω . For such a range of scale, these AVOs will come with a range switch to break down these ranges to manageable levels.

To measure current (Amps) we must break the circuit where we wish to measure and place the meter in series with the circuit – a terminal point is usually the most convenient, or across an open switch

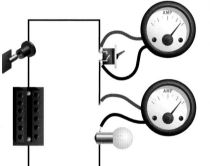

if there is one. When set to measure current, the meter must offer the least possible resistance, otherwise it would act as an obstruction to the current flow itself and register a false reading.

To measure voltage or voltage drop, the meter leads need to straddle the component being measured. Unlike measuring current, when measuring voltage the meter itself must offer the maximum resistance possible, since we don't want any of the current normally flowing through the component passing into the meter – yet again, this would yield a false reading.

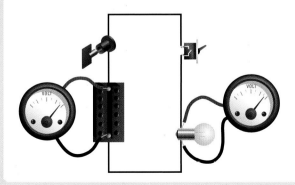

To measure the resistance of a circuit, the complete circuit must be isolated from all power and the meter leads used to close the circuit again.

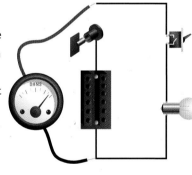

The meter introduces a small current into the circuit and measures how much resistance it meets – the result being ohms. To measure individual components – such as fuses – it's often necessary to disconnect the item from its neighbours. For portable AVO meters the introduced current usually comes from a 9V internal battery.

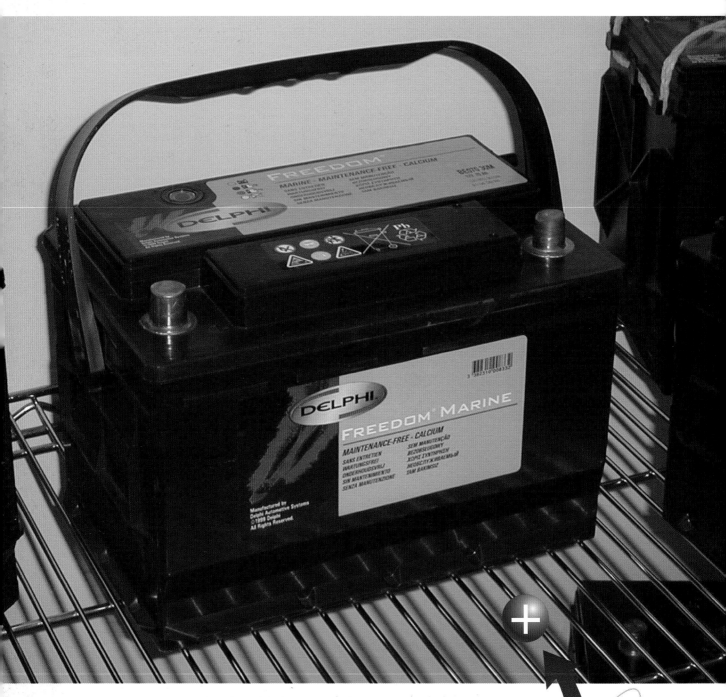

Primary & Secondary Cell

Lead-Acid Battery

Capacity

Deep Cycling

Monitoring

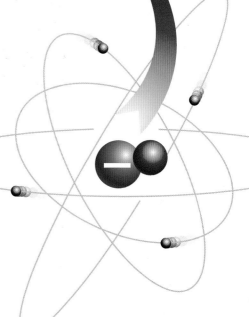

Batteries

2

The kind of batteries that power our torches, cameras and personal stereos are called 'primary cells'. Although fine for many purposes, such batteries are of little use for powering the electrical systems on a boat as they have a limited life and can't be recharged. What the boat owner needs are batteries that can be recharged repeatedly from some external source of power – perhaps from the mains when lying alongside, the engine when motoring, or by converting energy from the sun or wind. Batteries of this type are called 'secondary cells' and the most common of these are the lead-acid batteries we use to start our cars.

Most secondary cells are heavy and physically robust, but electrically they're quite fragile. To ensure a long and reliable life they must be treated and maintained properly – and here some knowledge of how they work helps a lot.

Primary Cell

A very simple experiment demonstrates how primary cells work. Obtain two strips of dissimilar metals – perhaps copper and zinc. Next fill a glass dish with salted water and put the strips into it so that they don't actually touch. Now take a voltmeter and connect

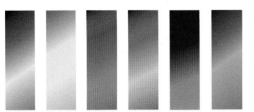

it across the strips and you'll see that the meter is registering a small voltage. The size of this voltage depends on the particular choice of differing metals paired off at the time.

If both or either one of the metals are lifted out of the water the voltmeter will register zero. When immersed again, the voltage and current is resumed. This shows that the salty water is a vital part of our battery. It's called the **ELECTROLYTE** and, without it, the battery wouldn't work.

But all primary cells have a serious drawback. If you were to leave our experiment running long enough you would find that one of the metals – the zinc in our case – would visibly corrode by the electrical (more correctly called 'galvanic') action. Once the zinc had become depleted, all activity would cease and the battery would be effectively dead. It seems, therefore, that the current has been created at the physical expense of the zinc.

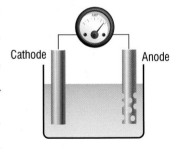

Cathode · · · · Anode

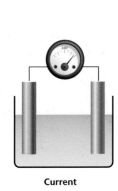

Current

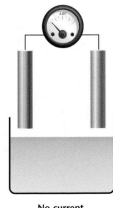

No current

Galvanic Series

If our little primary cell experiment was conducted with a wider selection of metals, then the voltages created would form the basis of a Galvanic Series Table. The range of potential voltage is quite small: from +0.2V for graphite to -1.6V for magnesium. Take any two and immerse them in an electrolyte and the current produced will have a voltage equal to the difference between them. Our copper and zinc strips would have generated 0.6V. But if we had substituted mild steel for the copper, we would only have seen 0.3V.

In any galvanic combination of metals, one of the two will assume the 'anodic' (least noble) role while the other will be 'cathodic' (most noble). The metal which is least noble will always take on the anodic role and will corrode as a consequence.

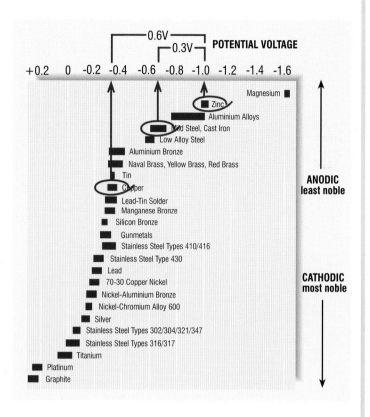

Galvanic Action

An interesting example of a primary cell is used in the lights fitted to life jackets and sometimes on life rafts. In these, all the components necessary to make the cell active are in place – except for the electrolyte. In its dry condition the cell is dormant. But if a crew member falls over the side with such a cell fitted to his life jacket, sea water (the electrolyte) floods into the cell which very quickly springs to life. The current thus generated illuminates a bulb, hopefully to lead rescuers back to the MOB.

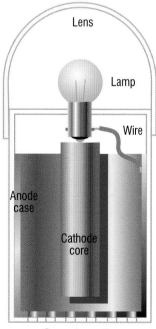

Left, a sea water-activated primary cell illuminating an MOB beacon. To the right, a cross-section of such a beacon. The life of the cell, and indeed the duration the beacon lamp will burn, will depend upon the quantity of anode material in the cell.

Secondary Cell

Another experiment, in some ways similar to the first, serves to show how a rechargeable battery works. Again, we use the same glass dish but this time partially filled with dilute sulphuric acid (battery acid). We also need a couple of lengths of plumbers' lead solder as well as some torch batteries. With the solder immersed in the acid and the batteries connected in series, you'll soon see one of the lead solder electrodes start to turn brown and bubbles rising from the acid. After half-an-hour or so, disconnect the torch batteries and connect a voltmeter across the electrodes. You should find that a healthy voltage will be registered.

Here in its most basic form we have a secondary cell. It differs from the primary cell in two ways. Firstly, we've used two similar metals instead of two dissimilar metals and, secondly, the electrolyte has changed from salt water to dilute sulphuric acid. What we've made is a device that can be charged from a flat state and then discharged without one of the electrodes being destroyed – in other words a rechargeable battery. Clearly, the function of the cell is due to an electro-chemical process between the electrolyte and the electrode material which can be reversed. This particular secondary cell simulates perhaps the most common type of battery used on boats today – that being the wet (flooded) lead-acid cell. So, how does it work?

Interestingly, when fully charged a wet acid cell will produce a voltage of about 2V no matter how large the electrodes.

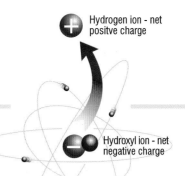

Hydrogen ion - net positve charge

Hydroxyl ion - net negative charge

Electrolyte

Whether a primary or secondary cell, the electrodes must be immersed in an electrolyte which forms a conductive lake between them. And the most important property of the electrolyte is that it be rich in ions. So, what are they? Page 8 describes the electrical and atomic balance of a molecule. But, should an atom leave a molecule, both the atom and the remaining molecule will be electrically imbalanced, either positively or negatively. These fragments are called ions. Water (H_2O) can contain ions where a single hydrogen atom leaves to form a hydrogen ion. Now deficient of an orbiting electron, it assumes a net positive charge. And, of course, the remnants of the water molecule are left with too many electrons and become a negatively charged hydroxyl ion. These ions occur naturally in water, but many more are produced when an electric current passes through it. The battery acid in a wet-acid cell consists of sulphuric acid and water. In its concentrated form, sulphuric acid is a non-conductor of electricity, but when diluted with water it breaks down the acid (H_2SO_4 – in which each molecule consists of two hydrogen atoms, one of sulphur and four of oxygen) into positively charged hydrogen ions and negatively charged sulphate ions. Because of their electrical imbalance, they make ideal current carriers, so resulting in a rich conductive solution where the higher the acid content the better the conductivity.

Discharging

In its fully charged state, a lead-acid cell will have its negative plate of lead and its positive plate of lead oxide both immersed in a rich solution of sulphuric acid. With the cell discharging, the positive plate will begin to lose its brown discolouration as the oxide on the surface is being removed with the help of the hydrogen in the electrolyte. The hydrogen is attracted to the positive plate where it combines with the oxide to form water. At the same time the sulphur in the acid pours into both negative and positive plates, combining with the lead to form the compound lead sulphate. So, quite a lot is going on. To recap: the acid content of the electrolyte is invading the plates while water is also being made – thereby, of course, diluting and weakening the acidity of the electrolyte. Both plates are being transformed into lead sulphate which, as they become similar in composition, lose their voltage potential as the cell becomes flat. The cell is discharged when it has no more lead oxide or acid in the electrolyte.

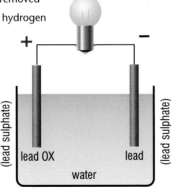

Charging

When it's recharging the process reverses. The sulphate content emerges from both the plates and goes back into solution in the electrolyte, thus converting the plates to lead. In doing so it strengthens the acidity of the electrolyte. The action of the charging current through the electrolyte separates the water into hydrogen and oxygen through a process known as hydrolysis. Both the hydrogen and oxygen then go into solution in the electrolyte, leaving the oxygen free to recombine with the lead of the positive plate to form a brown lead oxide. Meanwhile the hydrogen helps draw the sulphate out of the negative plate, contributing to its conversion to lead and the slight bubble activity is a mild venting of hydrogen gas. The cell is considered fully charged when no more lead at the positive electrode can be converted to lead oxide and when all the sulphate is driven out of the negative electrode. At this point the electrodes are as dissimilar as they can be and they have the maximum voltage potential between them.

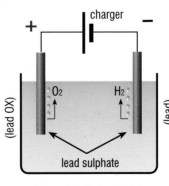

Lead-Acid Battery

With each cell producing 2V each, it follows that six of them must be stacked in series to give us the 12V we need for our boat electrics. In commercially made batteries the electrodes are made up of plates with separators between them. The plates are in the form of grids, filled with a soft lead paste. This allows the electrolyte to really soak into the paste where it can act upon the largest surface area possible. The separators are absorbent pads which perform several functions: they keep the plates apart while conducting the current through the electrolyte, and also help prevent the soft paste falling out of the grids – which could happen through vibration or any rough handling.

When you buy a lead-acid battery, it will be close to fully charged with the battery acid at its proper level. Or at the very least, the battery will be sold dry with the positive plates preloaded with lead oxide. The cocktail of diluted sulphuric acid would be added at home and then the battery would be charged.

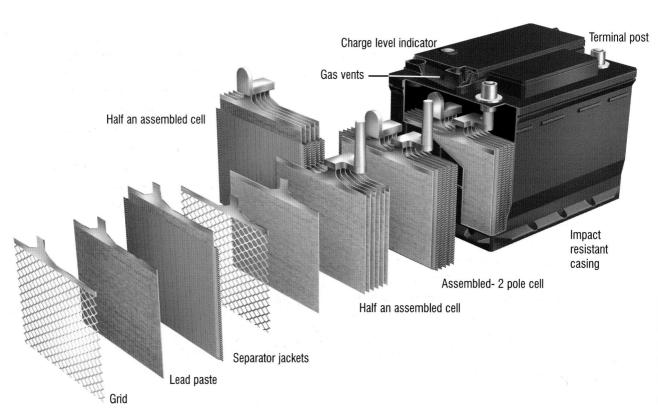

Charge level indicator

Terminal post

Gas vents

Half an assembled cell

Impact resistant casing

Assembled- 2 pole cell

Half an assembled cell

Separator jackets

Lead paste

Grid

Overcharging

NO SMOKING RISK OF EXPLOSION

This does several things, none of them good for the battery. With the battery fully charged, the positive plates are oxidised lead. Further charging will over oxidise the lead, making it weaker and more brittle – rather like the flaking rust you see on badly oxidised steel. Then there's our old friend hydrolysis, which separates the water molecules into hydrogen and oxygen when a current passes through the electrolyte. This is a quite normal part of the cell's regeneration process but, once the job is done, the hydrolysis continues superfluously. Electricians call it 'boiling'. Oxygen and hydrogen are given off from the plates and are vented to the open air. The electrolyte is being lost. More significantly, the combination of hydrogen and oxygen forms a potentially explosive mixture, which explains why battery compartments should be well ventilated and have warnings against smoking in the immediate area.

Over-Discharging

As soon as discharging starts, so too does the formation of sulphate on the plate surfaces. As the discharging continues, the formed sulphate acts as a barrier between the electrolyte and the plate material. This slows the reaction and limits the current that can be drawn. Over-discharging turns the initially soft sulphate deposits into larger, harder crystals which, once crystallised, are difficult to reconvert. This reduces both the battery's charging capability and its capacity. The condition is known as 'sulphation' and can only be reversed – and only to some extent – by subjecting the battery to a vigorous overcharge which, hopefully, will break up and dislodge the crystals (see page 67). However, this is a quick fix solution that can't be repeated too often since it's very hard on the battery, contributing to an early demise.

It should be pointed out that a battery can quite easily reach an over-discharged condition without even a load or a charge applied to it. Lead-acid batteries do self drain (see Efficiency) and can suffer the consequences of sulphation in the process. This explains why batteries should not be left idle and require regular exercise.

Battery Efficiency

We will all very soon realise that batteries aren't very efficient bits of kit. To avoid damaging them it's important not to discharge them below 50% of their capacity. And when you charge them you find they come rapidly to about 80% capacity, after which their internal resistance makes it progressively more difficult. This last problem can be overcome with modern boost chargers or regulators to squeeze the charge to perhaps 95%, but that still only leaves a useful operational window of between 35 and 45% – in the worst case hardly more than a third of what you might have thought you had. Also, in overcoming the internal resistance, alternators must put more energy into a battery than comes out. For every ampere hour consumed, the battery must first receive 1.4Ah of charging. The penalty factor of 1.4 makes a lead-acid battery a little over 70% efficient.

Other factors nibble away at the efficiency. A fully charged battery holds a considerable electrochemical force ready to supply current to anything that might bridge the gap between the electrodes. Any impurities in the electrolyte or dirt on the case could provide a productive path in parallel to the terminals through which the battery can self-discharge. Keeping the case nice and clean and the electrolyte topped up with distilled water will minimise these losses.

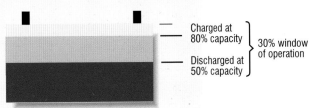

Charged at 80% capacity
Discharged at 50% capacity
30% window of operation

Capacity

When we purchase a battery we need a unit to measure the amount of electricity it will provide. One would think that batteries are sold like most electrical consumables by the KW or KWh, but power in KW is a product of both voltage and current. Since the voltage of a battery remains pretty much constant at 12V throughout its performance, it leaves the current as the only variable. This is fine, for what we really want to know is how long the battery can deliver and sustain a nominal size of current before it reaches its 'voltage end point'.

A typical boat battery will be labelled '12V 100Ah', so what does this mean for your money? Well this battery will provide 1 Amp of current for a period of 100 hours at a constant output of 12V throughout. Alternatively it would provide 2 Amps of current over 50 hours or 5 Amps over 20 hours at 12V and so on. We can see from each example that the product of the hours and the current gives us 100Ah.

The 10 hour rating stemmed from the early automobile days when it was thought to be a reasonable output for a car which travelled through traffic but was never on the road for more than 10 hours. Such a battery could have supported the load for that time under a dynamo which turned over slowly and put very little back into the battery.

When it comes to DEEP CYCLING a battery, there's an additional rating we need to understand. In the United States a battery's capacity is rated over a required discharge period of 20 hours, which means that a 100Ah battery will provide a steady current of 5 Amps for 20 hours. However, some batteries from the UK are rated over 10 hours, though recently some are adopting the American system. For the yacht owner, the 20 hour rating seems more logical as it's closer to the sort of time that typically elapses between when the engine is turned off and until it's turned on again.

The time over which a battery is rated often isn't identified. It's therefore important that you ask before buying, because there's a sizable difference between a 100Ah battery rated at either 20 hours or 10 hours.

Capacity against Discharge Rate

I f our 100Ah (20-hour rate) battery is subjected to different discharge rates, other facts become clear. At modest charge and discharge rates the voltage drops are small. With time on its side, the inner parts of the plates are accessed slowly enough for there to be not much in the way of internal resistance. Consequently, the battery will yield pretty much the same number of ampere-hours as were put in. But when the discharge is high, the lead sulphate forms quickly and blocks access to the material within the plates, slowing the reaction and restricting the current. The internal resistance rises and the **VOLTAGE END POINT** is reached sooner, leaving only a fraction of its capacity as being discharged.

So, the conclusion is that if we measure current against hours we will get the most ampere-hours for the lightest current drawn. To put it another way, the price you pay for high rates of discharge is an apparent reduction in battery capacity. Interestingly, if we discharge it at less than the 20 hour rate we actually get more ampere-hours than the rating would suggest. It seems, therefore, that discharging a battery either side of the nominal ampere-hour rating has significant implications for a battery's capacity.

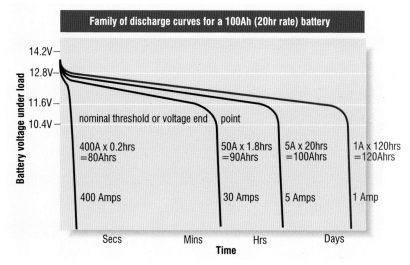

Voltage End Point

T he time it takes for a battery to discharge depends on its capacity, and the manner in which it does so can be shown graphically. The voltage at the terminals of a fully charged battery will be about 13.8V, which will fall to 12.8V almost immediately a load is connected to it. With further use the voltage will fall gradually to 11.6V when, to be on the safe side, the battery should be considered to be discharged. Although it may be thought still capable of further work (however feeble), to discharge the battery further would do it serious damage. The plates will become severely sulphated and the stronger cell plates may drive the weaker one into reverse polarity.

Below 11.6V the voltage drops steeply. When it reaches 10.4V it hits what is known as the 'nominal threshold' or 'end point' – and you should definitely go no further.

The rate at which the current is drawn obviously speeds the process, and it's helpful to plot this as shown on the graph.

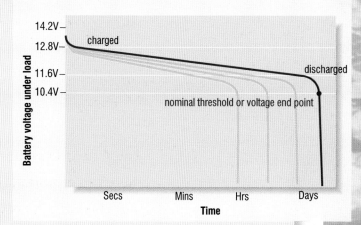

Deep Cycling

To deeply discharge a battery is to risk damaging it. To do so repeatedly is even more likely to kill it off. Yet at least some capacity for 'deep cycling' is exactly what you want on a boat, where there are often long gaps between opportunities to recharge. Think of a night passage under sail with your navigation lights ablaze, the instruments all in use, the radar turning and your autopilot working hard – perhaps as much as a 100Ah without the chance of a top up.

To deep cycle a battery means to bring it as near to exhaustion as you dare and without taking the voltage below its end point. This shouldn't be confused with what occurs when you start the engine or use the windlass – where in both cases the voltage could fall alarmingly. This isn't deep cycling but a short burst of high electrical consumption that causes the voltage to dip. As we shall see, the battery will achieve some form of RECOVERY. Deep cycling involves a steady draw of current over time, allowing the active material to be used up deep within the lead plates, leaving little scope for recovery. Some batteries will tolerate deep cycling much better than others – a subject we shall touch on later in this chapter.

Recovery

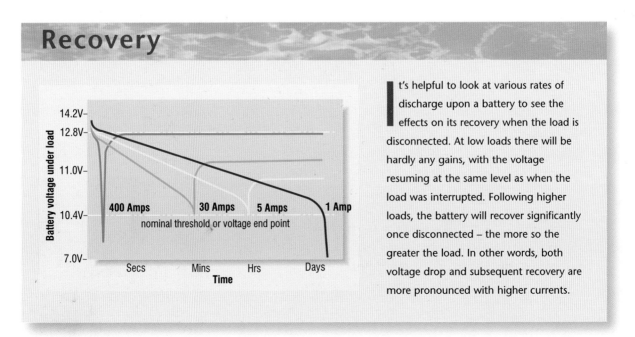

It's helpful to look at various rates of discharge upon a battery to see the effects on its recovery when the load is disconnected. At low loads there will be hardly any gains, with the voltage resuming at the same level as when the load was interrupted. Following higher loads, the battery will recover significantly once disconnected – the more so the greater the load. In other words, both voltage drop and subsequent recovery are more pronounced with higher currents.

Ampere-Hour Meters

It would be possible to keep a mental track as to the amount of charge put into the batteries and the energy consumed to give a figure of the charge remaining, but understandably this would be a headache to keep on top of and would more than likely lead to inaccuracies.

Thankfully, there are electronic devices available that can do the job for us. These are called ampere-hour meters and they're connected to the battery as shown below. By way of a shunt they are able to keep tabs on the current going in and out of the batteries, while the voltage connection monitors the battery terminal voltage. All it then takes is a little clever computing and an array of information can be displayed.

Some ampere-hour meters can monitor two separate battery banks at the same time through the connection of a horse-shoe shunt, (see P. 72).

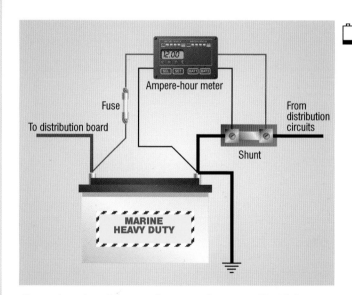

Ampere-hour meter

Fuse

To distribution board

From distribution circuits

Shunt

MARINE HEAVY DUTY

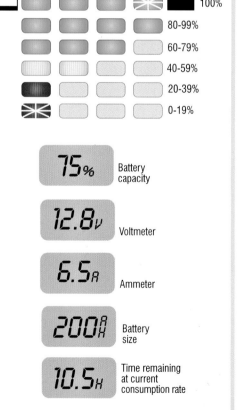

100%
80-99%
60-79%
40-59%
20-39%
0-19%

75% Battery capacity

12.8v Voltmeter

6.5A Ammeter

200A Battery size

10.5H Time remaining at current consumption rate

Above, shows how the ampere-hour meter is connected in relation to the battery and the shunt. The LED light sequence shown to the right provides the owner with an at-a-glance display of the battery's charging or discharging progress. Generally speaking, as soon as an amber light shows then the battery should be charged. Any red condition should be avoided for the battery is now under risk of sulphation.

To the right, is a selection of LCD displays on the meter giving an impressive range of information.

See Appendix C for Shunt

Battery Sizing

Before buying batteries, each owner should subject their boat to an ELECTRICAL AUDIT to replicate a representative lifestyle and consumption of power. This of course is essential if you're designing the whole electrical system from scratch, and it's also not a bad idea to do it every so often with existing installations, whose demands tend to grow as new bits and pieces are fitted over the years.

Electrical Audit

As a working example, let's look at a typical mid-size cruising yacht. We start our audit by making a list of every item that consumes electricity and record its power rating – usually marked or etched on it somewhere, expressed in watts.

By dividing the power in watts by 12V we obtain the current rating in amps, which are easier for us to use as they lead directly to ampere-hours, the standard for defining a battery's storage capacity.

The next task is to estimate how many hours each item would be consuming current over an average day. From this you can derive ampere-hours and end up with a table as shown below. The end product of the table will be a total consumption of electricity for this particular boat over an average day.

Having estimated the daily consumption, you're then in a position to decide how much battery capacity you need – remembering that the very maximum you can get out of it is about 45% of its rated capacity, and it's more likely to be 30-35% (See battery efficiency P. 23). So, for our specimen boat's daily consumption of 80Ah you could just about get away with a couple of 100Ah batteries, but only if you had a very efficient charging system and perhaps some supplementary sources of power such as a wind generator or solar panels. A total of 300Ah would be more sensible and would give you a buffer against the natural fall-off in capacity that occurs over the years.

And this is just for the domestic services. The engine would usually have its own dedicated starting battery. However, if you happen to have a single battery system (see P. 59) serving both the engine and domestics, then you would have to carefully consider both the battery's capacity and the best type to use, (see P. 30-31).

ITEM	CURRENT RATING (A)	HOURS RUNNING	AMP HOURS
CABIN LTS (3)	3	3	9
VHF	1	8	8
FRIDGE	5	8	40
INSTRUMENTS	0.5	8	4
RADIO/STEREO	1	4	4
AUTOPILOT	1.5	3	4.5
ANCHOR/NAV LTS	2	6	12

81.5 Ah

say **80 Ah**

Alternator Sizing

Once you've decided how much battery capacity you need, you then have to determine the output of the alternator that will charge them and at the same time support the domestic services. There are various rules of thumb, the most common being the one-fifth rule, where the charge output to battery capacity (C) would be C/5 or 20-25% of the total capacity – ie, for a 200Ah bank, you would need an alternator rated at 40-50A output. The reasoning is that charging a battery with a current greater than a fifth of its capacity might do it some harm. However, this rule is far from practical, and could leave you with chronically undercharged batteries. We need a more rational calculation that properly reflects each boat's specific demands.

A Balanced Input

Let's suppose we have 200Ah batteries, excluding the start battery which remains constantly topped up. Since we shouldn't discharge below 50%, we can divide the capacity by two – giving us a maximum discharge level of 100Ah. At this stage it's tempting to conclude that the alternator will need to be capable of putting 100Ah back in, but this isn't always possible. Most standard marine alternators have relatively crude regulators that produce a simple taper charge (see P.54) and will struggle to achieve more than about 80% of full charge. So, 100Ah x 80% = 80Ah is all we can reasonably expect, and the next question we need to ask ourselves is how quickly do we want it? The equation here is:

$$I = C/T$$

Where 'I' is the current from the alternator, 'C' is the net capacity and 'T' the charging time in hours.

The charging time (T) is up to the owner and the style of boating he or she undertakes. A motor boat skipper whose engine runs constantly when under way wouldn't object to an 8hr charge (and 80Ah/8hr = a measly 10A) while a bluewater cruiser might only want half-an-hour (80Ah/0.5hr = a whopping 160A, and quite absurd for it would certainly cook the batteries).

The middle way might be a weekend sailor who would run his engine for about an hour and a half a day, giving 80Ah/1.5 = 50A say. But, you also have to take into account the batteries' inefficiency – about 70% (see P. 23) which would mean 1.4Ah of charging for every single ampere-hour put in. So 50A x 1.4 = 70A, just to recharge our bank in one and a half hours. Unfortunately, we haven't done yet, for we must add on the current needed to support the ongoing electrical demands during this period, something you must estimate for yourself. If it were 10A then the total output we would be looking for would be 70A + 10A = 80A, which is more powerful than the C/5 rule would have come up with.

Now let's return to our bluewater cruiser, willing to run his engine for just half-an-hour a day. Assuming he also has fitted an 80A alternator, what can he do to make up the 50-60 or so ampere-hour shortfall? The most popular solution is to fit wind generators or solar panels or both. A typical well thought out installation might comfortably produce a combined charge current averaging 2A over a twenty-four hour period, contributing nearly 50Ah to the general good.

The Range of Batteries

The choice of battery is important, for most types are either intended for or are better in one application than another. They may look much the same on the outside but it's what happens inside that counts.

Lead-acid batteries will either be fully serviceable, like our car batteries, where we can check the electrolyte level and top up

Starting Batteries

These are designed to produce quick bursts of energy and are generally poor on capacity. They have a large number of thin plates which, collectively, give the surface area necessary to deliver the high current needed.

The plates are intentionally thin because it enables the electrolyte to gain active use of the plate material in the short time it takes to start the engine. The surface area minimises the density of current, thereby reducing both the sulphate thickness and the internal resistance.

If you were to use a starting battery for domestic purposes, you would run into problems because deep cycling would cause severe sulphation. And a sulphated starter battery would have trouble surviving. The thinner plates would probably warp and are generally more vulnerable to the stresses of deep discharging.

The almost certain outcome of this would be that small amounts of sulphate would be shed from the plates to lie at the bottom of the case. Before long this accumulation would short out the plates to render them useless.

Cycle life: very poor, though they're rarely deep cycled

The two sectioned batteries above show the contrast between the starting and the leisure battery, with regard to the plate thickness, and the number of plates.

To the right, a deep cycle battery.

Leisure Batteries

These have a greater tolerance of deep cycling, though they are often confused with 'traction' batteries when in fact 'semi traction' would be a more accurate description. A leisure battery would have thicker but fewer plates which contain more active plate material over a smaller area to provide higher capacity. The inner regions of the plates aren't so easily accessible to the electrolyte, so they're less suitable for producing high starting currents, though they are sometimes used for this purpose, particularly when two are connected in parallel to form a boat's main battery bank – but such an arrangement inevitably takes away some of the flexibility of a two battery system where each can be used independently.

Cycle life: 200-300

as necessary, or 'maintenance free' which are sealed or have a gel electrolyte. We should always be very cautious about combining different types of batteries in parallel. The charging requirement for one type may not suit the other and could cause poor performance or serious damage. Always check with the manufacturers.

Dual Purpose

These offer a compromise between starting and leisure batteries: high capacity, some tolerance of high current and deep discharge, and a fairly long life. They are ideal for a single battery system.
Cycle life: 200

Sealed - Low Maintenance

Although ideal for such craft as jet skis which get bounced around a lot, this type isn't really suitable for a cruising boat where it would inevitably be deep cycled. The problem arises if they're taken down to less than about 40% full charge, from where neither an alternator nor a mains charger can properly bring them back. They can also be damaged by overcharging. The gases released inside are normally reabsorbed, but if charged too vigorously they won't recombine and pressure will build in the casing to the point where the relief valve will blow and electrolyte will be lost. As it can't be replaced, the damage is permanent.
Cycle life: 200-300

HEAVY DUTY	656		DIN EQUIV	62514	
AMP HOUR	126	AH	CCA SAE	810	AMPS
OTHER EQUIV	356		IEC	545	AMPS
RESERVE CAPACITY	220	MINS	CCA DIN	490	AMPS

12 VOLTS — DANGER DANGER — SHIELD EYES PROTEGER LES YEUX — SULPHURIC ACID ACIDE SULFURIQUE — NO SMOKING NE PAS FUMER — GAS EXPLOSIVE GAZ EXPLOSIFS — KEEP AWAY FROM CHILDREN

Top, a battery marketed under the label 'Heavy Duty'. Above is a detail of the battery lid label. Its capacity is 126 ampere-hours, and its reserve capacity is 220 minutes, meaning that the battery will supply a constant 25 amps of current for 220 minutes at 80°F before dropping to 10.5V. Two measures of Cold Cranking Amps (CCA) are provided under two different measuring standards; the SAE and DIN. The CCA is the minimum amps a battery can supply for 30 secs at a temperature of 0°F before dropping to 7.2V, and as such the CCA is quite an important rating when using a battery in the UK climate.

To the left, a sealed lead-acid battery. The little round window just above the Delphi label is a charge level indicator which substitutes for the facility of a hydrometer.

Gel Cell

Often called 'dry cell' or 'valve regulated lead-acid' (VRLA) these batteries have an immobilised gel paste as an electrolyte, and are very popular because there can be no spillage. Like sealed low maintenance batteries, the gases are intended to recombine with the electrolyte, only venting through a safety valve if overcharged.
Gel cells have a good tolerance of deep cycling and their self-discharge rates are low so they will hold their charge over the winter months. They can also survive if left in a discharged state, so are well suited to the rigours of shipboard life. Unfortunately, they're not suited to conventional charging techniques and are intolerant of any overcharging or the sort of regimes imposed by some fast chargers, though they do have a higher charge acceptance rate than lead-acid batteries. They require specialist, controlled charging to prevent overheating, since the gel is a poor conductor of heat. They also have a higher internal resistance, so can't deliver or receive current at high levels – rendering them unsuitable for engine starting. Finally, there's the cost: about twice that of a conventional lead-acid battery.
Cycle life: 400-800

AGM

Standing for Absorbed Glass Mat, AGMs are the next generation of batteries, whose separators are of glassfibre mat, designed to suck up the electrolyte between the plates by capillary action. This allows the ions to disperse more easily into the plates than they would in a gel cell.
AGMs are both very tolerant of deep cycling and can deliver high starting currents, making them extremely versatile. As the electrolyte is completely absorbed in the matting, no leakage will occur, even if the case is cracked open.
Cycle life: 800–1000

Optima Spiral Cell

A variant of the AGM, each cell consists of one negative and one positive plate with a glass mat separator in between, all wound into a tight spiral. The spiralling provides a large surface area so the internal resistance is extremely low.
Other advantages are high mechanical strength, no plate shedding, a very low self-discharge rate (they claim they will hold a charge for up to 4 years) and no gassing unless overcharged.
Cycle life: 800-1000

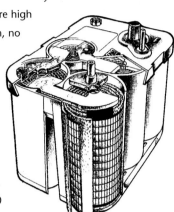

Traction Battery

Commonly found in such things as golf carts and milk floats, these batteries have tubular plates and are often sold as 2V units (requiring 6 to make up 12V). They're extremely robust and can be drained down to about 10% of full charge, making them ideal for a boat's domestic supply. However, they're useless for engine starting so it's essential to have a dedicated battery for this purpose.
Being almost indestructible, traction batteries are a very attractive option for bluewater cruisers.
Cycle life: 1000 and upwards

Carbon Fibre

This excellent conductor is used in the construction of the cell grids which hold and bind the paste together. The millions of strands of carbon vastly increase the surface area, enabling a fast and efficient charge. The fibres also act as a capillary pump to draw the acid deeper into the plate. This both reduces sulphation and gives more potential capacity.
Cycle life: 1000

Monitoring

It's obviously important to know how much electricity you have left. Unfortunately, taking a voltage reading at the terminals isn't a good gauge because the difference between a fully charged and an effectively fully discharged battery is very narrow; 12.8 Volts and 11.6 Volts respectively.

A better method is to measure the change in strength of the electrolyte as it loses acidity during discharging, and gains it again as it recharges. Pure sulphuric acid is 1.83 times the density of water – ie its specific gravity (SG) is 1.83. This can be measured with a hydrometer, which is basically a bellows syringe with a weighted float inside. A sample of the electrolyte is drawn up into the hydrometer and the SG is registered on the calibrations on the float. Fully charged the SG will be 1.26; discharged it will fall to 1.12.

Below is a table showing a battery's SG through the range of the battery's voltage and state of charge. The higher the SG the better, for the electrolyte holds more acid and is therefore more conductive. This explains why higher currents can be drawn from fully charged batteries. It also explains why you should never allow batteries to become completely flat. If they do, all the acidity may be expended, leaving plain water whose low conductivity may resist subsequent charging.

Above, a hydrometer and far left, a close up detail of the float graduations.

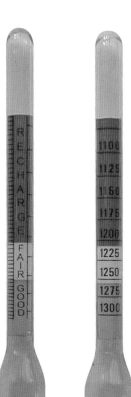

STATE OF CHARGE	SPECIFIC GRAVITY	VOLTAGE
100%	1.265	12.7
75%	1.225	12.4
50%	1.190	12.2
25%	1.155	12.0
DISCHARGED	1.120	11.9

Sulphation of batteries starts when SG falls below 1.225 or voltage measures less than 12.4V for a 12V battery

Start Your Engines

When did we last crack an engine? If you visited any of the major international boat shows, you'd be hard pressed to find any of the modern marine diesels even offering a hand start facility – the only emergency starter being a spare battery. Although we might question the wisdom of such developments, there's no doubt that this is the reality whether we like it or not. What this points to for the owner is the importance of knowing and understanding the engine electrics system.

Ground Return

Ignition Sequence

Sensors & Alarms

Gauges

The Return System

Back in the last chapter we saw the advantages of connecting multiple loads to a battery in parallel, via a two cable system of positive and negative leads. An engine wired up in such a way is termed an 'insulated two wire system'. These arrangements are not so common, but are essential for steel or aluminium boats. On the other hand, we can adopt the same principles of the simple parallel circuit but this time use the electrically conductive mass of the engine block as the negative side of the circuit. Such an arrangement is called a 'ground return system' and is by far the most common. It is also familiar to most of us in our cars.

Insulated Return

The most basic electrical circuit consists of a battery supplying a current to a load or appliance (shown for simplicity as a light bulb) distributed via two cables (Fig.1). It's worth noting that the conventional direction of current flows out from the +ve battery terminal, through the circuit and back to the -ve battery terminal. If we want to have a multitude of different loads from the battery then they're best connected in parallel to ensure an equal voltage across each appliance (Fig. 2). Now let's imagine a block of metal to represent an engine and the light bulbs as the engine's auxiliary appliances ie gauge sensors, alternators, and starter motors etc. Fig. 3 shows an arrangement where each appliance is insulated from the engine block and fed with its own dedicated cabling. This diagram represents the principles of an **INSULATED 2 WIRE SYSTEM** and is by far the least common, although it is the theoretically preferred one. However, this system is essential on steel or aluminium hulled boats where it is vital that the the engine block is insulated from the circuitry to prevent 'leakage currents' and the consequent ravages of electrolytic corrosion.

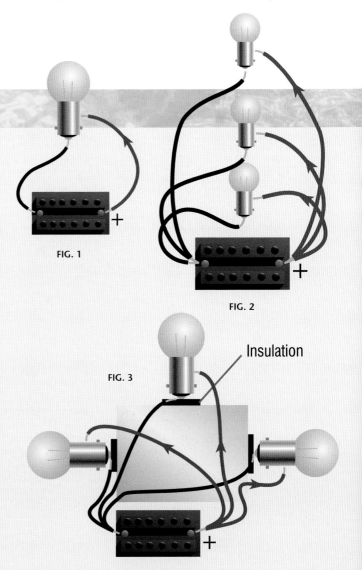

FIG. 1

FIG. 2

FIG. 3

Insulation

Engine System

Basically all the engine electrics in boats are in 2 sections – the Engine System and the Control Box System, (see Appendix A & B) either of which may or may not be supplied by the engine manufacturer. Nevertheless, once installed the two sections will be joined by a common 'loom' or 'harness' of cables with junction boxes at each end; each terminal or wire within the junction box will be either numbered or coded. Over the page, we can follow the functioning of the complete system as the ignition key is turned slowly through its various stages. Of course 'ignition key' is strictly incorrect when it comes to a compression ignition engine as opposed to a spark ignition engine, but the phrase is in common usage so we'll stick with it.

Ground Return

More often seen is the **GROUND RETURN SYSTEM** as shown in Fig.4. Here, one of the terminals of each appliance is connected to the engine block, thereby making use of the conductive nature of the engine to form one side of the electrical circuit – usually the negative side. Naturally this means you need just half the amount of cabling – not only simpler and less vulnerable, but also clearly offers the engine manufacturers and their customers some attractive

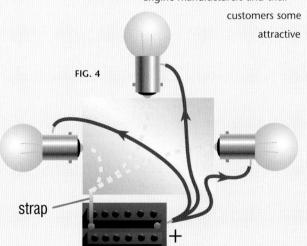

FIG. 4

strap

savings in cost. It is hardly surprising, therefore, that the majority of recreational marine engines employ this principle.

The negative side of the battery is connected to the engine block by a single heavy duty cable called a 'Strap' and the engine block itself, which acts like a sink of electrical current, is usually called the 'Ground'. This system is very similar to the one in your car and is pretty much standard throughout the whole of the automobile industry. Although some car manufacturers may use either the +ve or -ve polarity for the ground return, it's very unlikely that one will see a +ve ground return in the marine boating industry. The main reason for this is that it would make the engine block 'anodic' relative to everything else, making both the external and internal parts of the engine sacrificial to the more 'cathodic' parts (see AC Safety - Corrosion P.92). The terms 'earth' and 'ground' are often loosely used and if one isn't careful it could lead to confusion if not danger. To avoid any further misunderstanding, we shall use 'ground' when referring to DC circuits, and 'earth' for AC circuits.

Ignition - On

With the key resting in the 'Off' position, all the engine's circuits are dead. When turned to the 'On' position, light duty current from the +ve terminal **1** of the cranking battery (about 1 Amp) passes through the battery isolator switch **2** via line **3** to the +ve terminal of the starter solenoid **4** . Then through the ignition key **5** to power up the gauges **6** and their lamps where the needles of the gauges will be seen to flick over. The current from the gauge lamps is returned to ground via the tachometer return lead **7** which leads directly to the -ve terminal of the alternator which happens to be grounded to the engine block. From here the current is picked up by the -ve terminal of the starter motor, which is also grounded to the engine block, and the return current is finally carried to the battery by the strap cable **8**. As well as the gauge lamps, the voltmeter's return line also hitches a ride on the tachometer return lead to complete its circuit.

The currents from the temperature and pressure gauges **4** and **5** are returned to ground via a parallel path through their respective sensing lines **G1** & **G2**. The contacts of the lube oil alarm sender unit are normally closed at engine standstill so at this point the sender **9** assumes an alarm condition. This means that the circuit in line **W2** is closed enabling the buzzer alarm **10** to sound since the positive current from the buzzer is made to ground via diode **A**. This does annoy some people but it can be overcome by installing an optional timed alarm relay and making a slight adjustment in the wiring. This arrangement delays the activation of the alarm condition and allows enough time for the engine to start up and build up its oil pressure to its running level. Leaving the alarm system as is, however, does have the advantage of giving an audible warning that the ignition is 'On' and it also tests the lube oil alarm every time you start the engine. The charge light and its associated circuitry **11** plays an important part in supporting the alternator by confirming that a charging current is flowing between the alternator and the battery (see Getting Excited P.52), so expect to see the charge light come on. The circuit's return to ground is achieved via the alternator's voltage regulator.

See Appendix C for Solenoid and Appendix F for Diode

Ignition - Start

Once the ignition key is turned a stage further to the 'start' position then a more moderate current of about 10 amps flows down line **3** from the +ve terminal of the battery to the ignition switch **5** Through the position of the switch, the current is returned to ground via line **12** thus energising the solenoid coil **13** of the starter solenoid **4** The energised coil closes the relay to bridge the heavy duty terminals of the starter solenoid. It thereby allows the heavy duty current to flow to ground via the starter motor and the strap cable **8** before finally returning the current to the -ve terminal of the battery. The purpose of this relay is to relieve the engine control box circuitry of this viciously high starting current which could be in the order of 400 Amps. The engine should now crank over and start to pick up speed eventually to the start throttle setting. The lube oil pressure will build up sufficiently to open the sender alarm **9** contacts which shuts down the alarm circuitry **W2** and alarm **10** Once the engine has taken over the start sequence and is running, the ignition key **5** is released and it springs back to the 'on' position. Now the starter solenoid relay **13** is de-energised which opens the start motor circuit and stops the start motor. Also in the 'on' position, all the gauges, sensors and alarm senders remain connected. For engine control panels with pressure and temperature gauges it is more than likely that there will only be a common alarm shared by all the senders, so in the event of any alarm condition it requires a glance at the display gauges to establish which is at fault. However, for a panel without pressure and temperature gauges, these gauges will be replaced by individual warning lights – namely High Water Temperature, Low Lube Oil Pressure and Low/No battery charge. They will still share a common buzzer alarm as before.

There may be an occasion when, despite a healthy engine start, an alarm condition is still present. This condition is most likely to be a Low/No Battery Charge Alarm and this is merely a sign that the alternator hasn't developed sufficient output voltage yet, probably due to the alternator not being excited enough (see Getting Excited P.52). To resolve this, increase the engine rpm slowly until the lamp and the buzzer go out, then return the throttle to idle.

Ignition - Stop

To stop the engine, hold the 'stop' button **17** in against its spring until the engine comes to a standstill. Pressing this button energises a solenoid (line **17**) on the fuel pump which pulls the fuel throttle rack to stop the fuel to the pumps. Once again as the engine stops the lube oil pressure alarm at the sensor will switch in and sound the alarm; turning the ignition 'off' will silence the alarm and shut down all the electrical circuitry of the engine. Some engine panels have the engine stop sequence incorporated within the ignition key. To stop the engine, turn the ignition key to 'engine stop' – a position midway between 'on' and 'off'. The switch will energise the fuel pump solenoid as previously described. A problem associated with this arrangement is that it's very easy to leave the key in the 'stop' position once the engine has stopped and eventually the fuel pump solenoid will burn out.

The illustrations in Appendix B show in greater detail the electrical wiring behind a simple engine console panel, typical of say a 10 to 20HP engine. The fixed wiring at the back of the panel is fitted to provide a complete modular assembly, whereby the cluster of loom terminals from the engine's harness is simply plugged into their respective terminal slots on the panel. Similarly at the other end of the harness, the cluster of loom terminals plug into their respective terminal slots on the engine.

Sensors and Alarms

The size and power of an engine will often detrmine how elaborate its control panel will be. Whereas smaller engines might simply have visual or audible alarms to guard against overheating or loss of oil pressure, larger ones will have gauges which will tell you precisely how well the engine is performing. To do this, there must be some means of sensing temperature and pressure and getting the information back to whatever monitoring system happens to be installed.

The devices that pick up this information from the engine are called transducers and there are basically two types – sensors and senders. Sensors relay their information to the gauges, while senders relay their signals to activate the alarms. In some instances both the sensor and the sender can be combined in one unit to create a multi-functional transducer.

Some multi-cylinder high powered engines may have a gearbox interlock coupled into the starting circuit which will basically ensure that the engine can only be started with the gearbox selected in neutral. This will give the starter motor at least a decent chance of turning the engine over without overloading itself.

An oil sender to the left and a water temperature sensor to the right. The latter has only one terminal, indicating that this particular engine has no high temperature alarm for its cooling water.

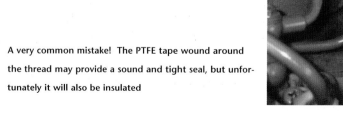

A very common mistake! The PTFE tape wound around the thread may provide a sound and tight seal, but unfortunately it will also be insulated

Pressure and Temperature

The cooling water temperature sensor (top) uses a thermistor pellet. This is a metal whose electrical resistance changes with variations in its surrounding temperature. If this device is coupled up to an electrical circuit, we can incorporate a gauge **14** to pick up any changes in the current as the thermistor alters its electrical resistance within the circuit. The only moving part, and hardly even that, is the bi-metallic switch for the high temperature alarm. This switch incorporates a strip of two dissimilar metals of different expansion rates bonded together in a pre-set bend shape. With changes in temperature the different rates of expansion cause the strip to bend. If this strip were incorporated into an electrical circuit one could see that this bending movement could act the part of an 'on/off' switch. Under normal engine running conditions these alarm contacts are normally open and closed on alarm.

The lube oil pressure sensor **16** (left) is basically a hydraulic/electrical device. The oil pressure is collected in a chamber and the pressure acts across a diaphragm. Any movement of the diaphragm is turned into a mechanical movement of levers where an electrical contact arm is swept across an electrical resistor coil called a variable resistor or rheostat. Again, if this device is coupled to an electrical circuit we can incorporate a gauge **15** to pick up any changes in the current as the rheostat alters its electrical resistance within the circuit.

The lube oil pressure alarm sender (bottom) is again a hydraulic/electrical device but simpler. The oil is collected in a chamber and lifts a piston against a very light spring. Once the piston is lifted from its seat it opens an electrical contact. A complete fall off in oil pressure (down to 0.1bar) will re-seat the piston and close the contacts to make the alarm circuit. Some engine manufacturers may combine the pressure sensor and alarm sender in one unit. This device will be much the same as the oil pressure sensing unit and the low pressure alarm contacts will be incorporated within the mechanic linkage between the diaphragm and the rheostat.

Both the sensing and sending devices have common ground terminals, that being the threaded stud fasteners that screw into the engine block. The threads of these are carefully and deliberately tapered to ensure a tight seal and good electrical contact. There have been many boat owners who have wound plumber's thread tape on the threaded fasteners, thinking this provide a secure gasket. Probably true, but then they wonder why the alarms are a bit quiet and there are no readings on the gauges. Also there's a risk that the thread tape will tear and breakup on screwing in and may enter the oil system to block vital oil ways.

Should any one of the three alarm sensing devices go into an alarm condition, say the high water temperature, for example, then the +ve current from line **6** is free to flow to ground through the buzzer, via its respective diode **B** and through the closed switch **W1**. The same principle applies for a low oil pressure condition, except that the buzzer circuit flows via diode **A** instead. The purpose of these diodes is to prevent one alarm circuit from back feeding into another alarm circuit.

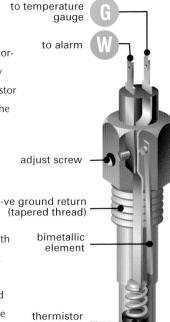

to temperature gauge **G**

to alarm **W**

adjust screw

-ve ground return (tapered thread)

bimetallic element

thermistor pellet

to pressure gauge **G**

variable resistor

diaphragm

pressure chamber

-ve ground return (tapered thread)

to alarm **W**

insulated cylinder

contact piston

pressure chamber

-ve ground return (tapered thread)

See Appendix C for Thermistor, Bi-metallic Switch, Variable Resistor and Rheostat

FOLD OUT APPENDIX A

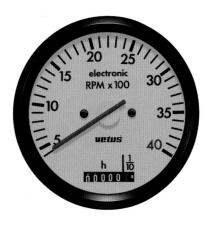

Gauges

Most engine gauges are mounted on the control panel well away from the engine– usually in the cockpit somewhere. Nearly all the gauges today are electrical and are driven by signals from transducers which transfer the particular mediums being monitored into electrical signals, that are much easier to convey around the boat. Some control panels will have alarms rather

Engine Tachometers

Tachometers can monitor both the engine speed and engine hours, where the rpm is registered via an analogue dial and the hours via rotating tumble wheels. The type of tachometer shown below derives a pulse from the alternator's AC winding, either from the star point or any one of the unrectified phases. The alternator output signal frequency is directly proportional to engine speed, and this signal frequency is applied to an electrical gauge calibrated to represent engine rpm. To rotate the tumble wheels in accordance with real time, the pulse signals from the alternator need to be rectified. If not, then the tumble wheels would rotate in accordance with the engine speed and not its running time. So, this rectified DC signal is then applied to a small DC motor which rotates the digital tumble wheels which are calibrated to rotate and represent real time. Depending on the alternator make, the terminals for the tacho will be labelled either W, AC, X, AC TAP or R. At the gauge, it is not unusual to find its negative or ground return fed directly back to the alternator's ground terminal. Any other ground returns from the engine console, may also hitch a ride on the tachometer's ground return lead.

Other sources of signals for tachometers could include an electromagnetic pick-up. Here, a magnet located on the rim of the engine

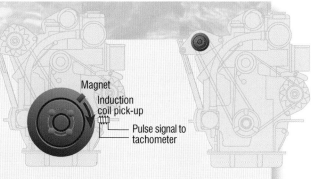

flywheel passes an inductor coil (which is fixed to the engine block) every revolution. As before, the induced pulse signals are then applied to an electrical gauge on the panel. Another source of signal could be from a small engine-driven generator whose output would be proportional to engine speed. This output would then be applied to a voltmeter calibrated in rpm.

Some panels may not have a tachometer, but simply an engine hours meter. This gauge could be a DC motor geared down and calibrated to represent time. Industry is now moving towards digital liquid crystal displays, and in this case the hours gauge would be an LCD clock counter - a notable disadvantage of which is that it often has no reset function. The source of signal for this meter would be from the alternator auxiliary output which, of course, will supply a DC signal for as long as the engine is running.

Some tachometers and hours gauges are connected across the ignition positive and the negative. Although deemed more practical in some respects, there's the possibility of false readings, since if the engine was stationary and the ignition was left on, the meter would still be keeping count.

than gauges for the engine's vital systems. Those that have gauges are most likely to have one or more of the following: oil pressure, water temperature, engine rpm and/or engine hours, engine exhaust temperature, fuel level, charge light/battery light and voltmeter.

Pressure and Temperature

The gauges for the oil pressure and water temperature are most likely to be voltmeters as well, with similar terminals as below, but don't be surprised to come across a slightly more sophisticated gauge with additional terminals and inputs from the mains. These gauges are called 'crossed coil measuring devices'. On the electrical side, the sensors at the oil pressure and water temperature heads are variable resistance devices so it does make sense to use accurate measuring devices to pick up the subsequent changes in current. The intricacies of these meters are not important, but basically they work by having two coils set at right angles to each other, sharing a common axis within a common magnetic field. One coil is fed from the mains supply and tends the pointer towards infinity, whilst the other coil which is fed from the sensing circuit, tends the pointer towards zero, hence a balance is struck. Variations in the voltages generated affect both coils in the same proportion and, as opposed to a normal voltmeter, there are no return springs or damping coils for the meter to work against. This means that despite any variations in the battery voltage, the gauge reading will remain the same. This makes cross coil meters very accurate and, more importantly, consistent throughout their operating range, as opposed to normal single coil meters which have to work against a return spring which is notorious for a phenomenon called 'range error'. It is very easy to tell if you have a cross coil device, since when the engine is stopped and the ignition is turned off, the gauge does not return to zero.

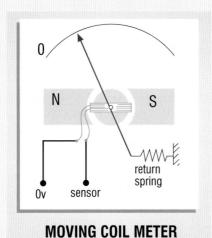

MOVING COIL METER

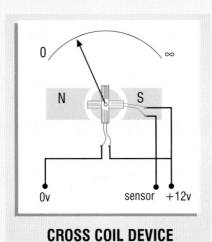

CROSS COIL DEVICE

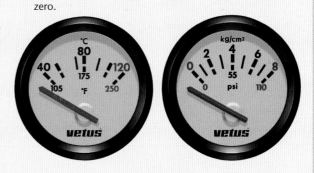

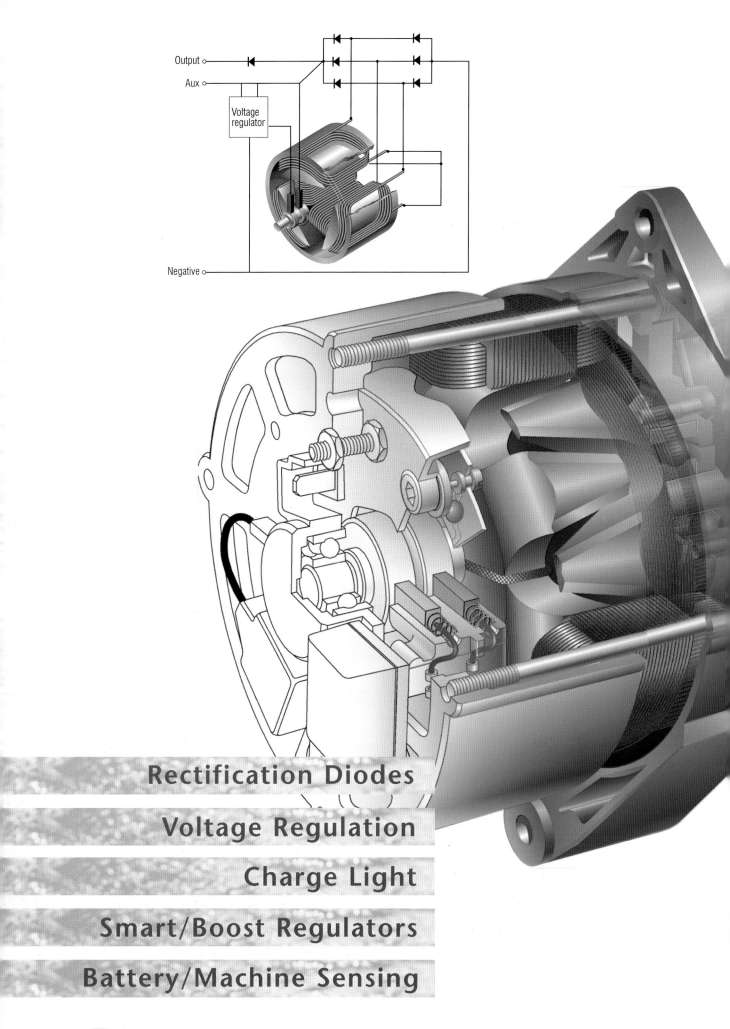

Output

Aux

Voltage regulator

Negative

Rectification Diodes

Voltage Regulation

Charge Light

Smart/Boost Regulators

Battery/Machine Sensing

Alternators

On most boats, a running engine provides a useful and attractive means of motive power to drive an electrical generator to supply the boat's services and at the same time charge the batteries. But oddly, as some might think, alternators produce an AC current when a boat's electrical power is stored and supplied in a DC medium. In this chapter we look at the reasons why this should be so and also how the electricity is created and controlled.

Electricity generated by motion involves conductors moving within the influence of magnetic fields. The electricity from the power stations to the front door of our homes is sourced the same way. Many of us are also familiar with the bicycle generator which spins against the wheel as you pedal along; it is little known perhaps, that the simple bicycle generator is in fact an alternator. Early cruising boats used dynamos which are DC machines and provide obvious benefits for DC boats, but the advantages were outweighed by their high maintenance and limited scope. Today, alternators are commonplace, bringing the benefits of simplicity, cost effectiveness, a wide scope of operation and ease of control.

4

Generating Current

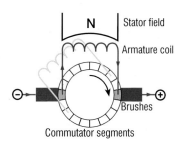

N Stator field

Armature coil

Brushes

Commutator segments

The commutator provides a means of collecting generated current from a spinning rotor. It is a cylinder of copper segments in which each opposing pair of segments is connected to a respective armature coil. The generating circuit is made when each opposing pair of segments contacts and passes through a pair of carbon brushes, rather like a rotating electrical switch. From the diagram, there is no continuity for the grey coil, but there is continuity in the commutator for the red coil.

The Dynamo

Dynamos are DC motors working 'the other way round'.

If you apply a current to an electric motor, it turns. And if you spin an electric motor by some external means, it produces a current. Dynamos were very popular some years ago, but then along came solid state electronics which made alternators more attractive.

Dynamos at first produce AC current, which is then converted to DC by the commutator. Since the commutator is a major source of wear, dynamos need more maintenance and are also susceptible to failure. To minimise this wear, dynamo drive pulleys are geared down so that they only run at engine speed. And since they're very sensitive to speed, this creates problems when trying to charge batteries with the engine ticking over.

The Alternator

Whereas the dynamo draws its current from its rotating armature, the alternator generates it within the stationary cage ring or stator coils in almost exactly the same manner as the simple bicycle generator. This makes transferring the generated current to the main terminals a much simpler matter, dispensing with all the complexities of a commutator.

To induce current in the stator, the rotor must be magnetised.

For reasons explained later, permanent magnets aren't suitable in a marine alternator; instead electromagnets are used.

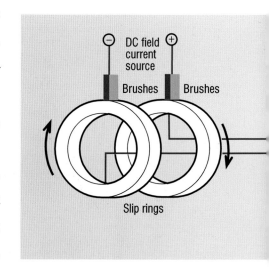

DC field current source

Brushes   Brushes

Slip rings

The Bicycle Generator

This is as simple as it gets! These devices are actually alternators, producing an AC current to light up the headlamp. A fixed stationary coil is influenced by a spinning magnet. The faster you pedal, the faster the magnet spins and the greater the current produced.

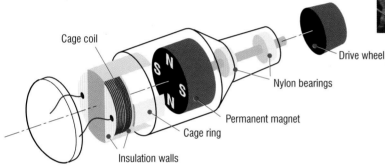

Cage coil

Drive wheel

Nylon bearings

Permanent magnet

Cage ring

Insulation walls

It can be seen from the photo that only a single cable is used to distribute the power to the bicycle's lights, with the returning current being conducted by the bicycle's frame – what's known as a ground return system (see last chapter). Most marine engine-mounted alternators are connected in the same way.

These are powered by a steady DC current, fed to the rotor coils via slip rings which are lightly loaded and are generally much more reliable than commutators. Thanks to this inherent robustness, alternator pulleys can be geared up to 2:1 or even 3:1

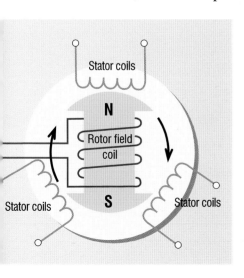

Stator coils

N

Rotor field coil

S

Stator coils

Stator coils

of engine speed (a bicycle alternator is about 25:1!) which means that alternators remain effective over the whole range of engine revs. The rated output of the alternator, engine speed and the drive pulley ratios is a carefully balanced affair and is something that should be left to qualified hands and not be experimented with.

AC/DC

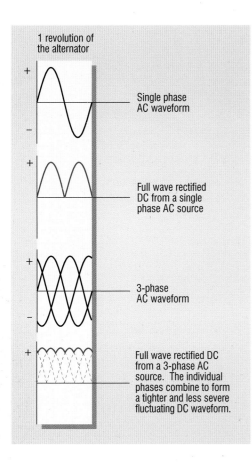

1 revolution of the alternator

Single phase AC waveform

Full wave rectified DC from a single phase AC source

3-phase AC waveform

Full wave rectified DC from a 3-phase AC source. The individual phases combine to form a tighter and less severe fluctuating DC waveform.

Initially the generated output produced from the stator coils is a smooth sinusoidal pulse of electricity called alternating current (AC). To be of any use and to be compatible in a boat with a DC system the alternating current must be converted or rectified to DC. This is achieved through the use of solid state electronics – namely diodes. A nest of usually between 6 and 9 diodes are used, achieving a result known as full wave rectification. Cheaper alternators are 'single phase' which rectifies to a very crude DC with peaks and troughs from 0V to +12V twice every revolution. This may not be good enough for sensitive electronic gear. More sophisticated alternators have a three-phase output which, when rectified, produces a much smoother and cleaner DC current, where the severe peaks and troughs associated with single phase full wave rectification are substantially smoothed out.

See Appendix F for Diodes

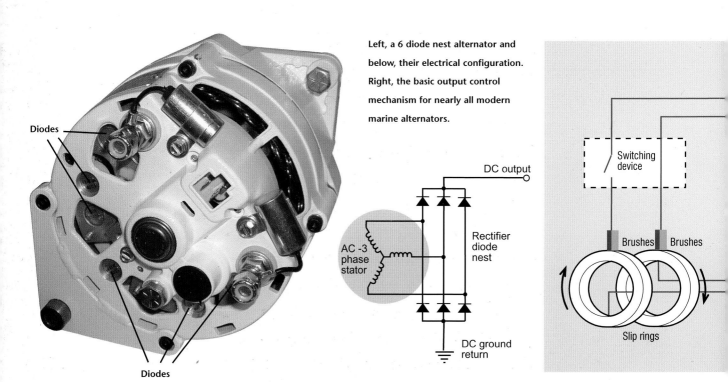

Left, a 6 diode nest alternator and below, their electrical configuration. Right, the basic output control mechanism for nearly all modern marine alternators.

Diodes

Diodes

DC output

AC -3 phase stator

Rectifier diode nest

DC ground return

Switching device

Brushes   Brushes

Slip rings

Alternator Control

To be useful for charging the batteries and generally serving the boat's electrical system, an alternator's output must be held within fairly tight tolerances. An alternator generating either side of those tolerances could damage both the batteries and the distribution services.

The elements that affect output are: (1) the speed (in rpm) of the machine; (2) the number of turns in the stator windings; and (3) the strength of the magnetic field. Of these, (1) is set by the manufacturer and is therefore beyond our control. Affecting any control by varying (2) is impracticable because the engine's primary purpose is propulsion. This leaves (3). By adjusting the current in the rotor field coil (called the field current) we can obtain the control we want since any changes in the rotor current influences the strength of the magnetic field by a directly proportional amount. Ingeniously, the current in the field coil is actually tapped off from the alternator's own DC output, thereby making it self-sufficient in generating its own magnetic field and, of course, its productive output.

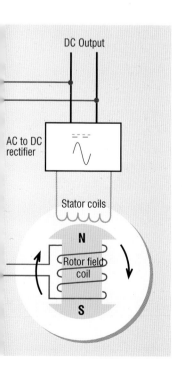

If we used permanent magnets to create the magnetic field, then the output voltage of the alternator would race away out of control as its speed builds up. Output would have to depend upon rotation speed and precise control would be impossible.

Having found a method of output control, we now need to find a means of controlling the output precisely. Having a switching device between the alternator's output and the field coil provides that means. Such a switching device is called a VOLTAGE REGULATOR.

Voltage Regulator

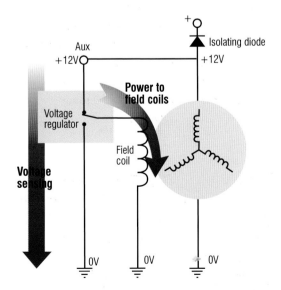

For a regulator to function, it must fulfil two criteria. First it must have real power to send to the field coils and second, it is essential that there is a direct connection between the regulator and the ground – after all, it's the voltage difference between these two points that the regulator is trying to set to 14.2V (The reason for this size of voltage is explained under charging on P. 58). Some regulators are electromechanical which stems from the dynamo days. Today though, most alternators employ solid state electronics, whose main components consist of transistors and Zener diodes. Such regulators usually form part of the alternator by being mounted at the back of it. Others are located away from the alternator and more often than not these regulators are likely to be 'optimum fast charge devices' or SMART REGULATORS: these are covered in more detail later. It isn't necessary to know how they work but it is vital to make sure that the regulator is connected properly to the alternator.

Most regulators on modern alternators are physically located on the back of the alternator itself (left). However, the electrical position of the regulator could either be before the field coil or after the field coil, and these are referred to as P-type or N-type regulators respectively, simply because they're either connected to the positive or negative side of the field coil. There's hardly any difference between the two as far as advantages or disadvantages are concerned, and these types of regulators should not be confused with NPN and PNP transistors. Generally P-type regulators are more common in the USA whereas the N-type is more common in Europe.

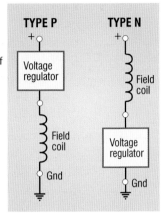

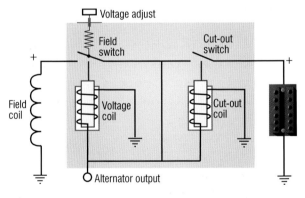

THE ELECTROMECHANICAL REGULATOR

The cut-out switch prevents the battery from draining current to the field coil when the alternator is stationary. As soon as the alternator generates current the cut-out coil closes the switch. A rising alternator output voltage will cause the voltage coil to work against the spring loaded adjustment setting. Once overcome, it will open the field coil switch, and as the alternator output voltage falls, the adjustment spring will overcome the voltage coil to close the field circuit switch.

Sense of Duty

For any standard regulator to work it must sense the alternator's output from somewhere. The sensing can be taken from one of two places and this is very much dependent upon the type of charging system employed. We can either sense the alternator's very own output or we can sense the battery voltage that the alternator is actually charging. Which one we use and why we use it becomes apparent when we look at charging systems, but for now we only need to be aware that this option exists and that the alternators are called 'machine sensed' and 'battery sensed' respectively. At first glance, when looking at the two types of alternator there would seem to be no difference between the two. However, the give away is at the back of the alternator, where the machine sensed type has three leads connected whereas the battery sensed type has four – the fourth obviously leading off to the battery.

MACHINE SENSING

The diagram below shows an alternator with an N-type regulator which is machine sensed. The regulator is sensing the voltage from the alternator's primary output to ground and since our example is an N-type regulator, the field coil is already receiving power from the output as well. If the sensing registers an output potential less than the pre-set voltage (about 14V) then the regulator will switch to close the field coil to ground and so establish a field current. The alternator's output should now develop and if the output potential climbs higher than the pre-set voltage (about 14.4V) then the regulator will switch to open the field coil to ground to deny the alternator any field current.

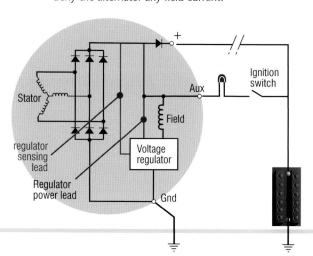

BATTERY SENSING

This diagram shows an alternator with an N-type regulator which is battery sensed. The sensing lead can be taken direct from the +ve terminal post of the battery, or alternatively it can be taken from the charge light side of the ignition switch to ensure that there is no drain from the battery once the engine is stopped. The other end of the sensing lead is taken to an additional terminal on the alternator, usually labelled S, IGN or B+, which goes directly to the sensing circuit of the regulator. The regulator senses the voltage from the battery to ground, and since our example is an N-type regulator the field coil is already receiving power from the auxiliary output. If the sensing registers an output

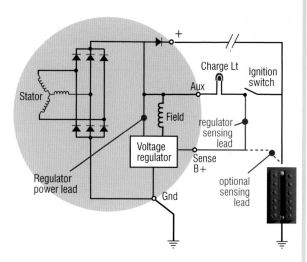

potential at the battery less than the pre-set voltage (about 14V) then the regulator will switch to close the field coil to ground and so establish a field current. The alternator's output should now develop, and if the output potential at the battery climbs higher than the pre-set voltage (about 14.4V) then the regulator will switch to open the field coil to ground to deny the alternator any field current.

Adverc's smart regulator which is branded as a battery management system.

Getting Excited

The self-sufficiency of alternators raises an interesting chicken-and-egg situation, in that they won't generate an output without field current and there won't be any field current until they produce an output. So how does an alternator start from scratch? Well, luckily, the field coils are wound around soft iron cores which primarily helps to concentrate the magnetic flux over the generating coils in the stator.

The secondary function of the soft iron is that the induced magnetism from the field coils remains in the cores after the alternator has stopped running. This phenomenon is called 'residual magnetism' and it does have a limited life. In most cases, through regular use, it's enough to generate an output which in turn will energise and excite the field coils and 'flash' the machine alive. An alternator that achieves this by itself is called 'self-excited'.

But if an alternator stands idle too long, the residual magnetism can degrade to the point where self-excitation is impossible. To guard against this, a branch feed (shown on the left), known as an excitation circuit, is bled off the engine's ignition circuit, and conveniently has the secondary function of confirming that a charging current is flowing from the alternator to the battery through the charge light.

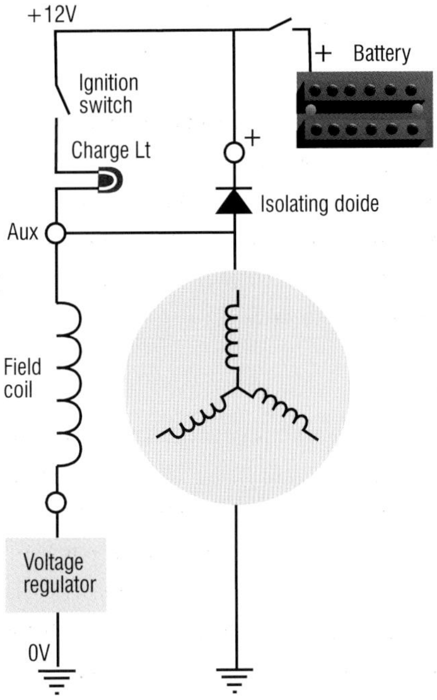

CHARGE LIGHT

When the ignition switch is closed, the 12V charge indicator light illuminates and brings the potential at the Aux terminal just above 0V, leaving just about 100mA of current from the cranking battery to energise the field coils and 'flash' up the alternator. Once the engine fires, the alternator starts generating, which in turn makes the voltage at the Aux terminal slightly higher than the battery voltage of 12V. Now, with such a meagre potential difference across the 12V lamp, the light goes out. There may be an occasion when, despite a healthy engine start, the charge light remains 'on'. This is merely a sign that the alternator hasn't developed sufficient output voltage yet – probably due to the alternator not being excited enough. To resolve this, increase the engine rpm slowly until the lamp goes out, then return the throttle to idle.

Death of Diodes

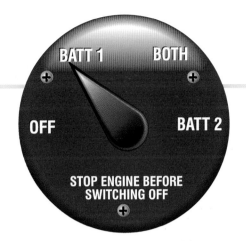

The most vulnerable part of the alternator is the nest of rectifying diodes. Electrically, these are sensitive components and are remarkably easy to destroy, and as a consequence its often the greatest cause behind alternator repairs. Most boat owners with manual charge control will be familiar with the warning notice shown on the right. The reason behind such a warning needs to explained in order to appreciate the damage that could follow, if it where ignored.

The effect of switching off a battery isolator or combination switch, is to open circuit the connection between the alternator output and the battery. If the alternator was generating at the time, then the current would have no where to go. However, it tries to keep flowing because of the inductance in the alternator output coils, and will develop what voltage it can in order to keep the output current flowing. The voltage spike created may be more than the regulator can handle and will be enough to pop off the diodes.

There are several ways of getting over this fairly common problem. Good selector switches have a make-before-break facility, without which, it would only take a slight pause in turning the selector switch to open circuit the alternator. Some selector switches go further by

making the switch part of the alternator's field circuit so that when the battery is isolated it also isolates the field current in the alternator, thereby preventing the alternator from generating that potentially damaging high current.

On diesel engines with manual stop controls you may also see another warning (above), cautioning you against switching the ignition key off with the engine

still turning. Again this is intended to protect the diodes, since the ignition switch and the charge light form part of the alternator's self-excitation system. Shutting it down is tantamount to running the alternator with a permanent magnet rotor, and accordingly the output voltage would race away.

SNUBBERS

By far the best fail safe of all, is to protect your alternator with a 'surge protection' or 'snubber' device should you inadvertently select the ignition to 'OFF' or isolate the battery with the engine running. A simple surge protector would be a Zener diode (sometimes referred to as a reverse avalanche diode) connected between the alternator output terminal and the ground. Any time the snubber senses an intense voltage peak it will dump the alternator output to ground, de-energising the field coils to stop any further output and so saving the diodes.

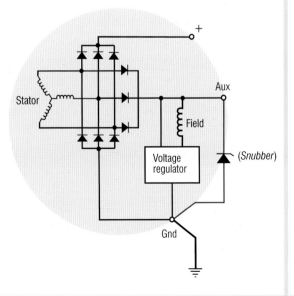

Smart Charging

It's in the nature of sailing boats that they don't spend a lot of time under power – for many, perhaps only when leaving or entering harbour. And the kind of charging system that comes as standard on most boats may simply not be up to the task of replacing the electricity used in its normal operation.

Cheap alternators and chargers will only ever achieve about 80% charge because of the battery's internal

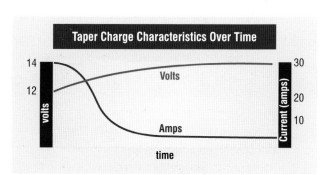

Taper Charge Characteristics Over Time

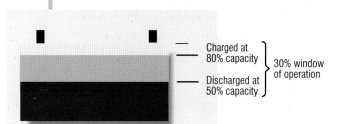

Charged at 80% capacity
Discharged at 50% capacity
} 30% window of operation

resistance. Then, of course, lead-acid batteries shouldn't be discharged below 50% without risking internal damage. This leaves a mere 30% of the capacity as a sustainable operational window – a meagre performance considering the demands we're likely to make upon it.

To overcome these problems we need to improve the alternator's output so as to make the most of those short periods under power. And the best way we can achieve this is to replace the relatively crude car type regulator with something more refined.

Let's look at what's available, first describing the simplest system before moving on to some more sophisticated solutions.

TAPER CHARGING

Lifted from automobile technology, this is the system we're most familiar with. The regulator is a simple device pre-set to an output level between 14 and 14.4 volts.

In its discharged state, the battery voltage will be low and will offer little resistance to the relatively high alternator output. Then, as the charging process advances, the battery voltage climbs and the chemical changes inside the battery increase its internal resistance and

opposes the incoming current. The progression is virtually linear until eventually the battery voltage and the alternator's output match each other and no further charge will be accepted, although a lead-acid battery will have reached only about 80% of its charge. This might be fine for a car, which charges continuously, but a boat could really use the full 100% – or at least as much as we can get of it. Clearly, we need a second stage of charging which will top the battery up.

3-STAGE CHARGING

Whereas the standard regulators are an integral part of the alternator, 3-stage regulators are mounted independently, either replacing the existing regulator or working in tandem with it. Some owners prefer the latter arrangement, since it provides a back-up if either of the two should fail.

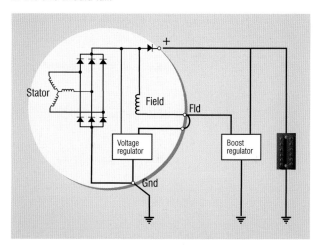

The 'Boost Regulator' offers its own advantages. When connected in tandem, it provides the additional advantage in that should any one regulator fail, the other will take over.

The three stages are called: BOOST or BULK, EQUALISE, and FLOAT or MAINTENANCE. During the boost phase the alternator's output current is maximised to bring the battery's voltage to a pre-set level of 14.2V as fast as possible.

Once there, the battery is 80% charged and the equalisation phase begins. The boost voltage is maintained but the charging current is allowed to relax over a pre-determined period until it's a gentle trickle charge. Eventually the battery is fully charged to 100% – any more and it would start to gas.

The flotation phase retains the trickle charge current but the voltage is allowed to fall to below 13.5V, just below the level where gassing can occur. If a significant load were applied to the battery at this point, its voltage would fall and the 3-stage regulator would immediately revert to a boost charge and start the cycle again.

A word of warning! Many of the most popular 3-stage regulators are made in the US and are designed to operate with P-type controlled regulators. Most European alternators are N-type controlled (see Regulators P.50). Some of these regulator units will have adjustment screws to control each of the three stages and, unless you're very sure of what you're doing, this is a job best left to the professionals.

CONSTANT CURRENT CHARGING

Again an externally fitted controlled device, these units always work in tandem with the original regulator, and start off by working rather like the 3-stage type. But this time, when the battery voltage reaches its initial target of 15.2V, the controller switches itself off, handing back the control to the alternator's own regulator, which continues to deliver its usual 14.4V.

The float stage is pretty much omitted, effectively making it a 2-stage regulator. The higher output rate can cause gassing in lead-acid batteries, so sealed or gel type batteries shouldn't be used.

PULSE CHARGING

This is yet another optimum charging system that works in tandem with the alternator's regulator. This type pulses the alternator output voltage for set intervals of time between 14 and 14.4V – not enough to gas. Typical settings are 5 minutes at 14V followed by 15 minutes at 14.4V. After about four repetitions, there's a 40 minute rest period at 14V.

The best controllers sense the battery temperature and adjust the charge accordingly – a higher voltage being optimum for lower temperatures and vice versa.

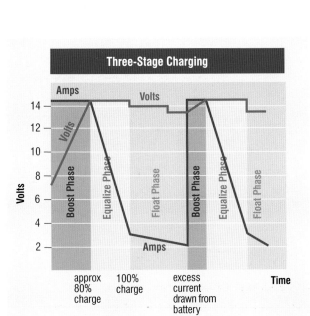

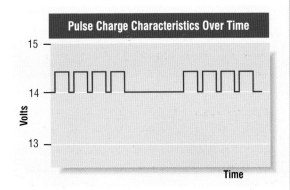

Charging Systems

An alternator, or any other generating device for that matter, has two distinct functions in a boat's electrical system; firstly it charges and replenishes the batteries and secondly it supplies electrical power to the boat's distribution system for general consumption. An alternator can simply be regarded as a pump, and as such needs to be incorporated into a system to achieve its distinctive tasks. One analogy similar to the functions of an alternator or charging system is our household water system, where the tank in the roof loft resembles the battery and the rising main water pressure from the pump station resembles the alternator.

5

Manual Charging

Split Relays

Split Diodes

Natural Energy

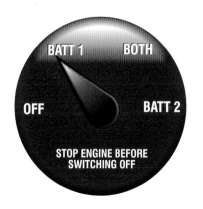

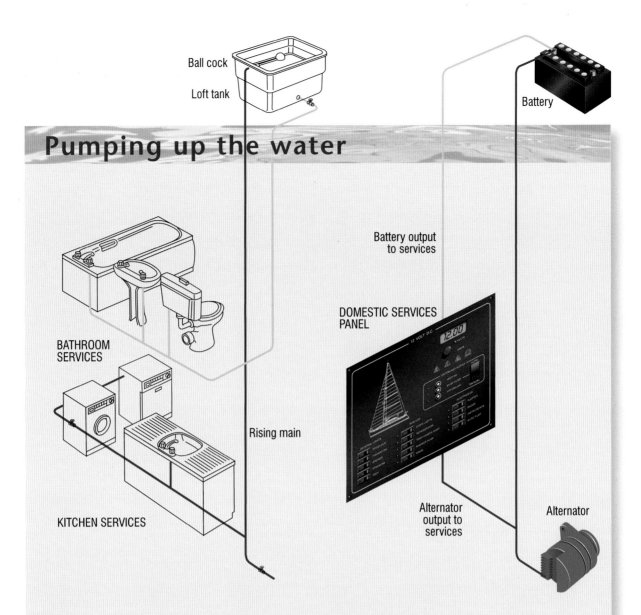

Ball cock

Loft tank

Battery

BATHROOM
SERVICES

Battery output
to services

DOMESTIC SERVICES
PANEL

Rising main

Alternator
output to
services

Alternator

KITCHEN SERVICES

I n order to understand what alternators do, it's helpful to abandon thoughts of electricity and think instead of the familiar territory of our domestic plumbing system at home. The typical house has a water tank in the loft which stores and maintains a head of water (a role similar to that of a battery). Being 20ft or so above ground level, the tank provides enough head to serve the various domestic outlets. The tank is connected to the rising main (the alternator) which is maintained at a certain pressure by the water company. The ball cock (the regulator) on the tank controls the flow of water to the tank and maintains its level.

When you run a tap, the tank starts to drain and the ball cock opens. To replenish the loss, the rising main must have a pressure greater than the 20ft head offered by the tank – any less and it will be overwhelmed by the head of water, and the tank will never refill, leaving the water to eventually drain away. While it's filling the tank, the rising main may also have to serve the

washing machine and dishwasher, both of which are often connected directly to it. Generally, for most of the time, demand from the water tank and the rising main will be slight to manageable. Then, when the demand becomes considerable, the delivery from the water main must be enough to refill the tank and support the supply to the domestic services. If the total demand from all outlets becomes greater or equal to the capability of the rising main, the tank will never refill or could drain entirely.

Much the same principles apply to an electrical alternator. If regulated to supply an output voltage of about 14V, it will have a 2V superiority over the battery's 12V and enough to charge a battery. We can think of this in much the same way as our head of water. But the quantity of electricity needed to satisfy demand is related to the alternator's current rating – its amperage – which is typically 50A to 120A. If you're running lots of appliances, you need a high output alternator.

Single Battery System

A typical alternator circuit in its simplest form is shown below. This is a single battery unit typical of a small inshore cruiser where the battery's duty covers both the engine starting and the domestic services. This is fine as long as the service requirements are small and there is little risk of exhausting the battery. The maindraw back of this system is that the domestic services will be subjected to large voltage drops when the heavy starter motor current is drawn from the battery. For the lights etc, this offers a tolerable inconvenience as they dim momentarily, but this may not be so tolerable for sensitive navigational equipment. The alternator output is usually taken to the heavy duty +ve terminal of the starter solenoid, and this point becomes the junction between the engine and domestic services. The

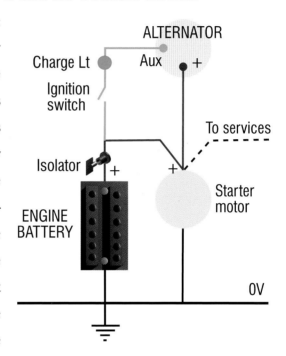

alternator in this particular single battery system is 'machine sensed'. This means that the regulator senses the alternator's own output as a means of controlling the voltage. Consequently a machine sensing regulator has no interest at all in what happens downstream of the alternator. We shall see later, that this type of sensing can be a penalty to an alternator.

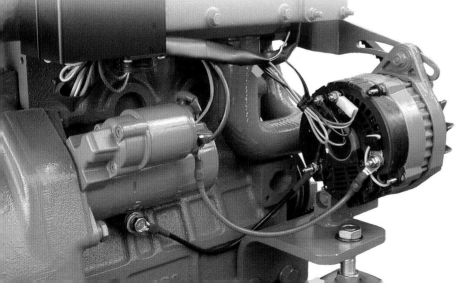

Multi-Battery Systems

For those having the space to accommodate them, a pair of batteries offers many advantages. Often, one battery is reserved for engine starting and the other, possibly a bank of batteries connected in parallel, serves the various electrical appliances fitted to the boat. Because there is no routine drain on the cranking battery, it should remain fully charged and so offer the security of being ready for use at any time.

Unfortunately, multi-battery installations bring their own charging problems. Obviously, both batteries must be charged from the same source (the alternator), and if they remain

Manual Switching

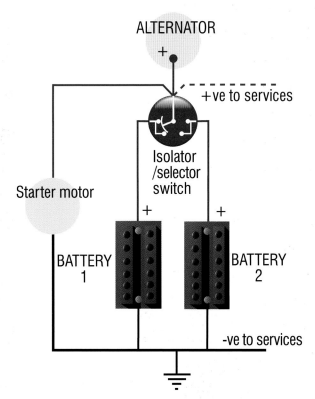

These switches are rotary devices with four positions: BATT 1, BATT 2, BOTH and OFF, and offer the simplest and cheapest solution to battery management. If, say, BATT 1 is selected, then all starting and other electrical services will be drawn from it and only that battery will be charged. The same reasoning applies for the positions BATT 2 and BOTH. Of course these are entirely manual operations and one must keep a mental track of knowing which battery is well up or well down in its charge. Also you have to remember to switch to the engine battery before starting the engine and switch back to the service battery after the engine has been running for a while. This arrangement is inherently prone to forgetfulness! Moreover, should you absentmindedly switch to OFF with the engine running, you could blow your alternator diodes. However many modern alternators have

If the alternator is stopped and both batteries are paralleled then the weaker battery will draw the stronger one down.

connected once the charging current has stopped, any imbalance in charge between the two will tend to level out, with the least charged battery pulling down the other. If one of the batteries were to become defective, it could deplete the healthier one to the point where both could be useless.

This problem is overcome by isolating the batteries in one of three ways.

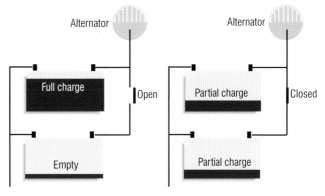

some defence in the form of an extra voltage limiting snubber or surge protection diode (see page 53). Incidentally, good quality selector switches are of a make-before-break type so that there's no battery disconnection as you switch between them. It's not uncommon to see both batteries left accidentally paralleled under load so that both batteries are flattened, leaving the obvious problem of no power to start the engine – quite a dangerous oversight!

Another potential trap can occur if you have difficulty starting an engine. After several attempts at cranking, your engine battery gives up the ghost. In desperation you switch to BOTH to reinforce its efforts. Bad move! Now a healthy battery finds itself coupled to one which is seriously depleted. Current rushes from one to the other and it is almost tantamount to a short circuit between the two batteries. If this is repeated often enough, you risk damage to the batteries.

So, is there a better way?

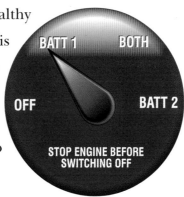

Split Charge Relay

Although rarely fitted nowadays, these devices were once very common. Usually wired into the engine ignition circuit, split charge relays isolate the batteries from each other when the alternator is not turning, but as soon as the engine fires up, they connect them together in parallel so that both may be charged.

Since there is a negligible voltage drop from the alternator to the batteries via this relay switch, it is usual for the alternator to be machine sensed and therefore have no cause to be interested in what's going on downstream of itself. However, very small voltage drops could creep in due to underrated terminals, cables and negative ground returns through the engine block.

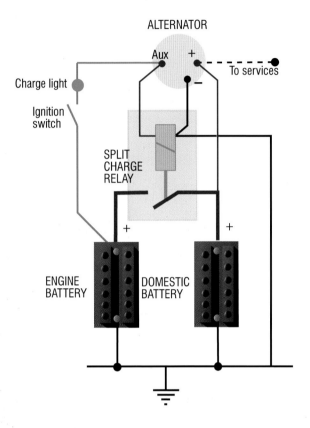

There's a widespread misconception that with a relay system the batteries become paralleled as soon as the ignition key is turned, and before the engine actually starts, thus risking the chance of weaker battery pulling the stronger battery down. If we look at FIG.1a with the ignition switch open there's no power to energise the relay. Once it is closed (FIG.1b) the 12V charge light will illuminate leaving a 0V potential at the alternator's Aux terminal. With a potential difference of 0V across the relay,

it's still not energised but the engine at least will be cranking over on its start up. As soon as the engine is up to speed the alternator will develop an output of 12V at its Aux terminal. Now we have a 12V potential across the relay which will pull in the relay switch to parallel the two batteries (FIG.1c). If, however, there was no charge light, then the aforementioned misconception would be valid, since there is no lamp to absorb the initial 12V drop when the ignition switch is closed.

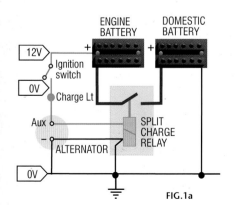

FIG.1a

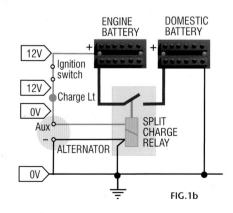

FIG.1b

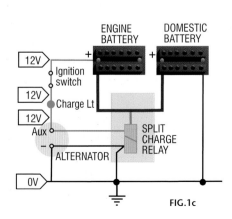

FIG.1c

Split Charge Diodes

These are now almost common practice because of their simplicity and reliability. The pair of diodes allows the charging current to flow freely to both batteries but does not allow current to flow between the two batteries. The battery with the lowest charge will be charged first, and it is possible to have different sizes of battery with each system as long as they have the same voltage. A manual selector switch can be incorporated into the battery circuit to allow the roles of the batteries to be changed and paralleled if need be. This is a good system offering flexibility and caters for most contingencies.

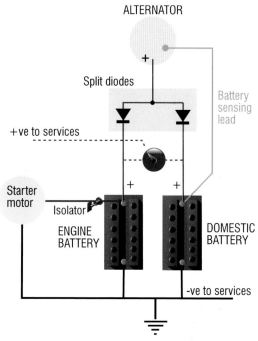

Despite all the praise diodes do come with a down side. In consuming current in the forward direction there is an inherent voltage drop of about 0.7 volts across the diode which is independent of the current it carries. This is a nuisance in the case of marginal charging when the whole of the charging output wouldn't be entering the battery. An alternator output of 14V would be lowered to 13.3V (14V-0.7V) by the time it reached the battery and the voltage head or potential may not be enough to charge the battery. A machine sensing regulator would be quite careless of this shortfall. So with this in mind, it becomes essential for the alternator and its regulator to be battery sensed so that any voltage drops are compensated for and the regulator will ensure that a full 14V reaches the battery. The domestic battery is usually the one chosen for sensing since it is generally the most used and the cranking battery is almost permanently fully charged.

Regardless of your charging system, a battery sensing regulator would be the best choice. In recalling the household water system (P. 58), the tank ball cock which 'regulates' the flow of water to the tank is, in a way, sensing the tank water level, in the same way that a battery sensing regulator senses the voltage level of the battery.

New alternatives to diode splitters are emerging on the market in the form of electronic charge splitters. These are electronic devices which emulate diodes; with these the voltage drop is insignificant, and machine sensed regulators would work fine.

Diodes run hot, and since diodes themselves are heat sensitive, they are usually mounted on a heat sink with cooling fins. The diodes should be located in a well ventilated space with an adequate updraft of air. A common mistake is to mount the diodes with the cooling fins in the horizontal plane, which creates numerous thermal pockets within the heat sink and rapidly shortens the life of the diodes through overheating. If diodes are mounted in the engine compartment, its wise to overrate them to compensate for engine heat.

A diode splitter: note the large heat sink surrounded by cooling fins.

Fooling the Senses

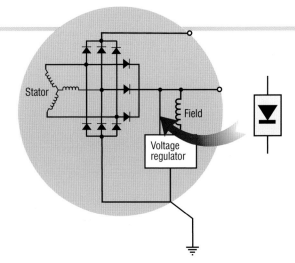

It sometimes happens that an owner wanting to install a split charge diode system discovers he has a machine sensed alternator. Now faced with the prospect of having to replace it, his enthusiasm dwindles. But all is not lost! By buying and fitting an inexpensive diode, he can play a trick on the alternator's regulator and fool it into thinking that it is sensing the battery potential.

The conversion itself is simple. Open the back of the alternator and fit a diode between the Auxiliary terminal and the regulator, as shown on the right.

Now let's see how this diode influences the charging circuit shown on the next page. The new diode will impose the same 0.7V voltage drop as the splitting diodes. And remember that the regulator is set to sense

A Compromise System - The Pathmaker Relay

This is a clever American device which might herald a renewed interest in split charge relays. As we've already dicussed, with a split charge diode system the alternator must be battery sensed, and usually it senses the domestic bank. This introduces a problem, because the domestic battery is often deep cycled while the engine start battery remains fully charged.

As soon as the engine starts, the regulator knows that the domestic bank is low and does what it can to replenish it. Unfortunately, the splitting diodes divide the charge current equally between the fully charged engine battery and the partially charged domestic one. The small amount drawn from the engine battery to fire up the engine is soon replaced and, thereafter, it would prefer to be left alone – a vain hope because the alternator is pursuing its own mission and continues to stuff electricity into both batteries. So, while the domestic bank is basking in all the attention it's getting, the engine battery is being assaulted, and possibly ruined, by a gross over-charge.

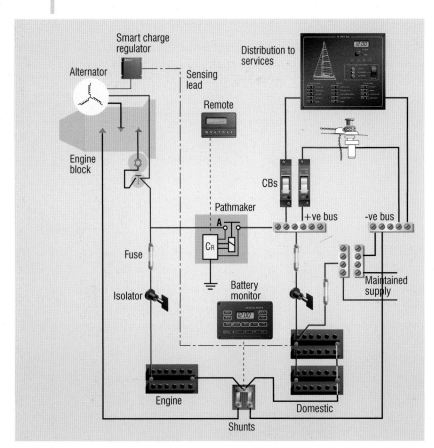

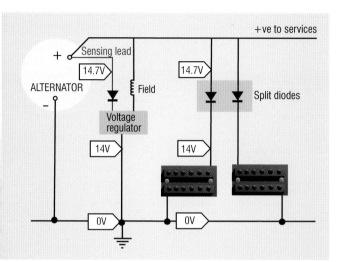

a 14V drop from the Auxiliary terminal to ground. At first, the alternator's output and Auxiliary terminal voltage will be 14V but, because of the new diode, the regulator will see only 13.3V which, of course, isn't enough. In response to this, the regulator boosts the voltage to 14.7V and as far as the regulator is concerned, it sees its pre-set 14V since the new diode has absorbed 0.7V from the 14.7V output. Thus far, the regulator is now happy with what it thinks it's doing. With the alternator now producing 14.7V, this means of course that 14.7V arrives at the diode splitters, which absorb that extra 0.7V leaving 14V to enter the batteries. Cheaper by far than a new alternator.

Which brings us to the Pathmaker, essentially a latching relay with a difference. Its main terminals are connected one to each bank, and the voltage at each terminal is continuously monitored with respect to ground. This makes it a voltage sensing relay which knows the state of both batteries and on its control box it allows you to set the voltage level for the relay to connect and disconnect the main contacts.

The alternator is still battery sensed and at the time of starting the engine, the Pathmaker contacts will be open. Indeed, when the engine is started the alternator will have no access to the domestic bank yet. The engine battery receives the whole output from the alternator and is quickly replenished. At that point the voltage potential at terminal A tells the Pathmaker's controller to energise the relay coil and pull in the main contacts. The two banks are now effectively in parallel and the domestic battery – being at a lower level of charge – will attract the most attention from the alternator, thus preventing the engine battery from being overcharged.

As soon as the engine stops, the voltage potential at the relay will no longer be at the alternator's output of 14V, but will drop to the 12V of the batteries. The Pathmaker responds by opening the main contacts, thereby isolating the banks from each other. If for any reason

the engine battery becomes flat, a switch on the Pathmaker overrides its normal function and connects the two banks together (for 5 minutes) to allow the domestic batteries to start the engine. So this relay plays the part of an emergency start facility as well.

The three adjustment screws on the panel board set the voltages at which the relay contacts make and break. The red switch is the manual override – used for an emergency parallel start of the engine.

Alternatives

Wind Generators

In a generally blustery climate such as ours, these provide a useful source of electrical power – a case of something for nothing once you've invested in the gear. Wind generators can operate day and night and typically produce 4 Amps in a 20 knot wind. This is fine for replenishing current used in normal domestic activities, but too high for topping up or trickle charging the batteries when the boat is left unattended. Therefore some form of regulation is required to control the output in relation to the battery voltage.

The generating machine will either be a dynamo or something resembling an alternator. The dynamos, of course, are DC motors working in reverse, so it's essential to have a diode on the output otherwise the battery that it is supposed to be charging will motorise the wind generator and you'll find yourself with an additional means of propulsion. The output of dynamos is usually twice that of the alternator type, and the commutator wear usually associated with dynamos is tolerable due to lower rotational speeds.

Alternator types simply resemble large bicycle generators (see P.47) and the generated current could be single or 3-phase before being fully rectified to DC current through a nest of diodes.

One common link between the dynamo and the alternator type is that they don't have field coils; they use permanent magnets to produce the fields instead.

Since for most of the time the wind generator will be idling away, it is only on the few occasions when the wind gets up, that the output becomes excessive and the need for a regulated output becomes necessary. It is therefore considered wasteful and complex to use any form of field current control to regulate the output, so other means and devices have to be adopted.

The most popular is the shunt regulator, which works on similar grounds to the transistorised voltage regulator. The bulk of the regulator's assembly consists of a heat sink and cooling fins. The regulator's transistors either switch the charging current to the battery or to a heavy ballast resistor housed in the heat sink which absorbs the energy as heat and then radiates it away. This heat can be considerable, and it has been known

for some boat owners to put it to good use by housing a shunt regulator inside a calorifier to provide free hot water. Shunt regulators have a low current rating and the strong output from an engine alternator can back feed into the regulator, burning it up and causing a severe fire risk. A protective diode is therefore required between the shunt and the battery – though if split charging through diodes this problem will be conveniently taken care of.

The shunt 'manual bypass switch' as shown in Appendix C is there so that the operator can periodically give the battery a full constant current charge to 'equalise' the battery cells. The purpose of this is to shed the crusts of sulphate that develop on the plates over the constant cycles of charging and discharging. The level of sulphation differs between each cell, and these slight differences in sulphation, and indeed sulphation generally by itself, brings the capacity of the battery down by as much as 20%.

So as a general rule of thumb, an 'equalisation' schedule of once every three months or so is recommended. The bypass switch is closed to finish off a charge for about 4 hours or when the battery voltage reaches 16V – whichever occurs first. This deliberate overcharging will force the sulphate to reconvert into active cell plate material and restore the cell to almost original condition. Gassing will occur, so re-topping the battery with distilled water will be required on completion. Also, equalisation should be done with the batteries isolated, since some sensitive electrical equipment won't like receiving 16V. This treatment is not necessary for gel type batteries.

Like engine driven alternators, the rated output of a wind generator versus the needs of the boat is a balanced affair. That also includes the size, pitch and diameter of the blades for the generator itself – so aviod swapping parts from other machines.

When purchasing a wind generator there must be a clear understanding between the dealer and the customer over the common ambiguity of what the actual rated power output is against wind speed. These values can be quoted as averages, peaks, maximums or even against nominated wind speeds - so watch out!

Most wind generators won't produce an output until a good 7 knot wind is blowing, and any generator sold on maximum output figures needs to be viewed with caution since it is likely that this output can only be achieved with a 30 knot wind speed. To be realistic you really need to know the output from a wind speed envelope of 10-20 knots.

Solar Panels

This is the silent alternative to wind generators, although to make it worthwhile you need quite a few panels, which can clutter up your boat. Solar panels are expensive and the large initial investment has to be weighed against the comparatively little power output in return. In terms of output per unit of money, on average, solar panels deliver less than wind generators, and even a thin veil of cloud cover can impair their efficiency by 70%.

Although the current produced is small, it is particularly useful in providing a trickle charge to maintain the batteries when the boat is left unattended. A panel will consist of a number of photovoltaic cells electrically arranged in series to provide a 14V output for a 12V rated system. A 2sq ft panel will produce about 6W of power – not a lot – but a vessel under sail will find that with a suitably sized panel there will be enough power, through a battery, for the electronic navigational equipment (but not an autopilot). Again, all but the tiniest panels need a regulator to prevent overcharging. Some wind generator regulators will also accept an input from solar panels, but the reverse isn't usually possible. Most dedicated solar panel regulators aren't up to the much higher currents obtained from wind generators. Some regulators have a separate output that can be used for circuits independent of the battery isolating switches – a sort of 'maintained supply' or an 'always on' system (see DC Distribution P.73). Such a system is ideal for when the vessel if left unattended since these circuits could look after automatic bilge pumps, intruder alarms and courtesy deck lights. If more than one panel is used then join all the leads to a common junction box, and from there the feed lead is taken to the batteries, via split diodes if desired, and the ground return lead is led to the -ve link of the DC circuit breaker board.

DC Distribution

A boat's DC electrical distribution system exists in a harsh environment, so it must be inherently robust. The wiring and the various devices must be electrically and mechanically sound, and also well protected so that they remain reliable over a period of years. Above all, the installation must be safe, for where there's electricity, there's also the risk of fire. Many people believe that because a voltage is low – usually 12V on all but the largest yachts – it must be safe. This just isn't so. The lower the voltage, the higher the current (in amperes) must be to do any given amount of work. And it's a large amount of amps which can cause a component to overheat, perhaps igniting surrounding materials.

Distribution Boards

CBs and Fuses

Maintained Supplies

Mast Wiring

Building on Basics

FIG.1

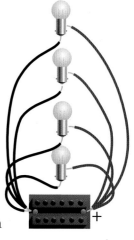

FIG.2

Let's start developing a typical boat circuit, starting with the simplest possible arrangement. Fig.1 shows a battery supplying current to a single load or appliance, here represented by a light bulb. It's worth noting that the current flows out from the +ve battery terminal, through the circuit, and back to the -ve battery terminal.

When more than one appliance is being served, these are best connected in parallel (Fig.2) so that each receives an equal voltage. But to wire a whole circuit that way would be both expensive and untidy. Luckily, we can start to make economies by sharing some of the return leads (Fig.3), bringing them together with the help of junction boxes. In this type of system the returning wire gets larger as you approach the battery. This is because it must handle the collective current of the various circuits feeding into it – rather like minor roads joining a motorway, adding to its traffic flow.

For safety, we need an isolating switch on the

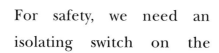

Junction boxes

Return leads

FIG.3

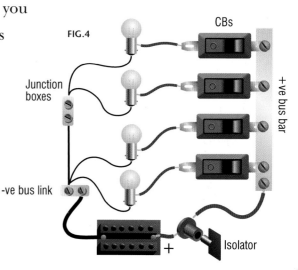

FIG.4

CBs

Junction boxes

+ve bus bar

-ve bus link

Isolator

Battery Banking

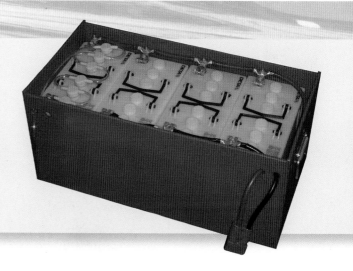

Most batteries on leisure craft are 12V, and the length of time that this output can be sustained to a given load is dependent upon its capacity, measured in ampere hours. The capacity can be a factor of battery size and type (lead-acid, Ni-Cad etc.) or by coupling up more batteries in parallel. This is known as a bank of batteries and, for all practical purposes, can be thought of as a single large battery having a capacity equal to the sum of them all.

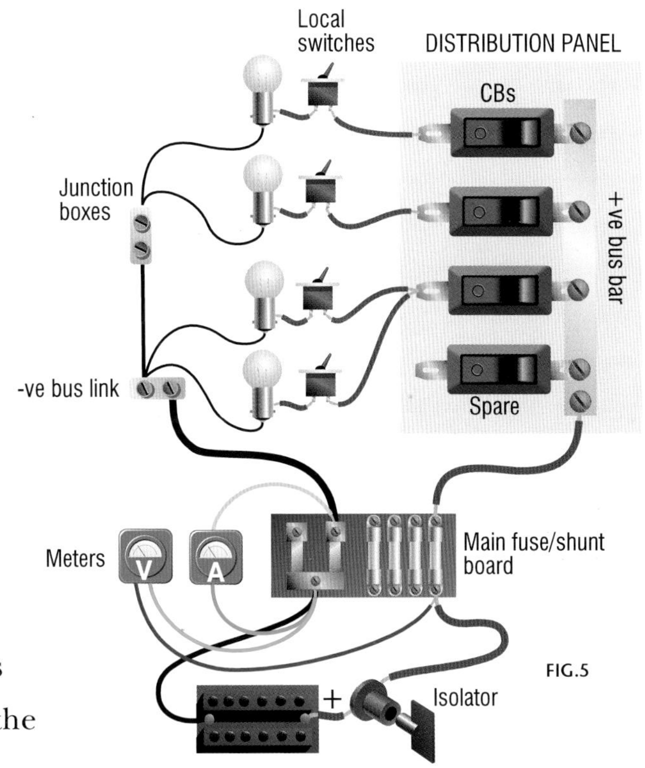

Local switches

DISTRIBUTION PANEL

CBs

+ve bus bar

Junction boxes

-ve bus link

Spare

Meters

V A

Main fuse/shunt board

Isolator

FIG.5

battery's positive side and fuses or circuit breakers (CBs) to protect the individual load circuits (Fig.4). By mounting the CBs on a common bus bar rail, we can use just one large cable to convey the power from the battery. This prevents cluttering up the battery +ve terminal and reduces the risk of any shorts across to the -ve terminal.

For convenience, the bus bar and protection devices are commonly mounted behind a labelled panel (Fig.5) from where the distribution system is controlled. Individual circuits reach out to serve the boat's appliances, each of which usually has its own on/off switch for local control. Primary circuit protection is provided by having an additional CB (or fuse) board between the battery and distribution panel – a very important function. This is a focal point where both sides of the primary circuit can be brought together, and forms a convenient platform to add monitoring instruments, such as a voltmeter and ammeter, the latter connected to a shunt. This primary protection platform is the most important of all the system's protection devices, since this is the last line of defence before reaching the batteries.

Battery Isolator

Each battery should have its own isolator mounted as close to its positive terminal as possible. Isolators come in two types: key operated or combined isolator/selector. The first of these will serve a single battery (or bank of them) and has the advantage that the key can be removed to disable that battery and deprive any would-be thief from having any useful power to the boat. As the name suggests, the isolator/selector type can choose between batteries,

isolate them, or connect them together to combine their power - perhaps to start the engine. As we saw in the last chapter, this often brings its own problems, because if you connect a fully charged battery to one that's flat, they could equalise so that neither has enough charge to be of much use. However, some form of emergency start facility is certainly desirable.

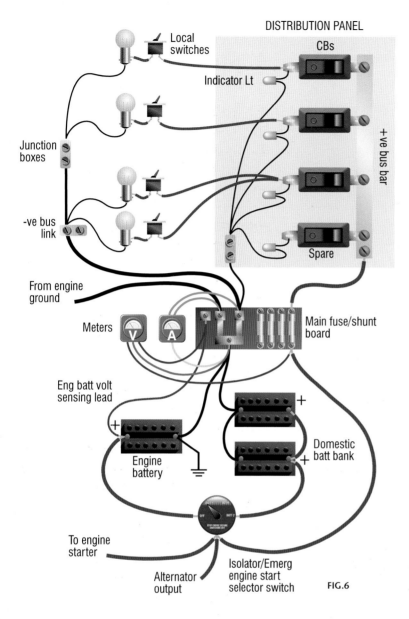

DISTRIBUTION PANEL

Local switches

CBs

Indicator Lt

+ve bus bar

Junction boxes

-ve bus link

From engine ground

Meters

Main fuse/shunt board

Eng batt volt sensing lead

Spare

Engine battery

Domestic batt bank

+

+

+

To engine starter

Alternator output

Isolator/Emerg engine start selector switch

FIG.6

With any multiple battery system offering a selection of battery services, as shown in Fig.6, there needs to be a common link between the negatives of all the batteries. You'll notice that the negative leads from both the engine and domestic batteries meet at one arm of the horseshoe shunt, thereby connecting them together. The voltmeter and ammeter also collect information from the shunt and, by using selector switches (not shown) on the gauges, you can monitor either system independently. As a further refinement, indicator lights or **LEDs** have been added to each of the local circuits to provide the owner with an instant indication of which circuits are on line.

Discrimination

This is the ability of a protection system to disconnect only faulty circuits while maintaining those that are sound. Discrimination is achieved by co-ordinating the current ratings and time settings of the CBs (or fuses) between the battery and load. As you can see from Fig. 6, there are two layers of protection in the distribution circuit (some sensitive equipment might be protected locally, making three). The devices nearest the load have the lowest current rating and the shortest operating time, with those nearest the battery having the highest current rating and longest operating time. If, say, a short circuit occurred at the top lamp, the fault current would be large enough to operate all the pro-

tection devices from battery to load. However, the lamp's CB (lowest current rating, shortest delay) should trip first to clear the fault and leave all other healthy circuits connected.

The last line of defence before reaching the batteries. The fuses leftmost on the board will have the highest current rating of all.

LED - see P.78

Don't switch off

This schematic circuit showing a second bus bar supplied by a separate feed, which adds the final touch. This is known as an 'always-on' or 'maintained supply' and is there to provide an uninterrupted supply of power for essential services - for example, automatic bilge pumps, emergency radio, etc.   As you can see, the supply is not isolated from the battery - it offers the shortest and most direct line from the battery. The maintained supply can be utilised to watch over a vessel which is left unattended. A security alarm system is one such utility, along with an automatic bilge pump which would be activated through a float switch (right) in the bilges.   However, these float switches need to be checked regularly since they do have a reputation for sticking, causing the bilge

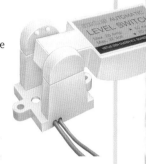

pump to run dry and overheat. A maintained supply is also utilised as a very important safety feature with regard to forced draft combustion heaters, widely found in the market place. These heaters are simply best described as miniature jet engines and it's absolutely vital that these heaters are allowed to go through their pre-programmed cooling down cycle before being shut down. For this reason you will see that the CBs in the maintained supply sub panel of the distribution board look different from those elsewhere on the board. These are 'resettable CBs' (bottom) which are manually closed and can only be opened through tripping. With regard to the heater, this ensures that no one can inadvertently interrupt the electrical supply to the heater whilst it is in the full flow of burning fuel. The heater can only be stopped at its dedicated controls which is assured of an uninterrupted electrical supply to complete the cooling down programme. In failing to provide this safety feature and if the electrical supply to the heater was suddenly interrupted, then unburned fuel would remain trapped in a very hot unventilated combustion chamber, leading to a very nasty explosion.

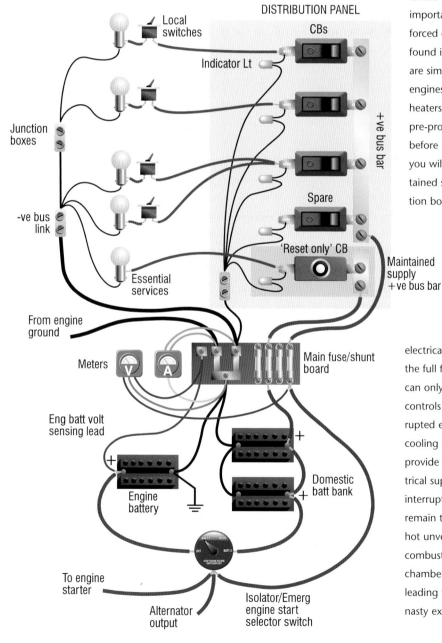

DISTRIBUTION PANEL

Local switches

CBs

Indicator Lt

+ve bus bar

Junction boxes

Spare

-ve bus link

'Reset only' CB

Essential services

Maintained supply +ve bus bar

From engine ground

Meters

Main fuse/shunt board

Eng batt volt sensing lead

Engine battery

Domestic batt bank

To engine starter

Alternator output

Isolator/Emerg engine start selector switch

To CB or not CB

That is the question – namely, fuses or circuit breakers? Which is best? And, if considering changing from one to another, could you, say, replace a 10A fuse with a CB of similar rating? The answer to this will help explain fundamental differences between the two.

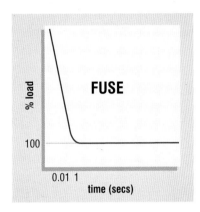

Some electrical loads take a large initial surge current when first switched on. Generally, a 10A 'slow-blow' fuse is intended to carry its rated current, plus any surge peaks which may normally occur in a circuit. The degree of these surge currents is a product of both the current and its slow-blow time, for example: 5 seconds at 2.5 x rated current and within 1 second at 3.5 x rated current. Taking a fuse above its rated value, like a surge current, does cause the fuse wire to burn and thereby weaken itself, and if repeated frequently the fuse will blow. To get over this problem, a 10A motor, for example, which has a starting current of 50A, may have to have a fuse far exceeding its normal 10A current rating to prevent the fuse from regularly blowing. This would leave the motor cables dangerously vulnerable to higher overload currents and overheating.

Fuses

Fuses can be found in 'quick-blow' and 'slow-blow' versions to suit different types of load. Due to their inherent simplicity, these are much cheaper than CBs but unfortunately offer few advantages - one being that a blown fuse obliges you to check the circuit, otherwise it will simply go again. A positive disadvantage is that they are open to abuse, for with a persistently blowing fuse, there is a great temptation to replace it with a higher rated fuse instead of its designed rating as a means of temporarily resolving the matter - many are guilty of this even with the full knowledge of the consequences. Another disadvantage, is that over time the electrically sensitive fuse wire will oxidise and just when you least need it the fuse will quite naturally break through old age so this should be borne in mind before searching for faults in the circuit.

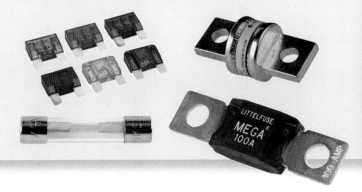

A circuit breaker, on the other hand, can do the same as the fuse above but this time through its robust design we can control the time delays incorporated within the CB to handle surge in current/overloads without damaging its own mechanisms. The 10A motor can now have a CB much closer to 10A and still be capable of carrying the 50A starting current.

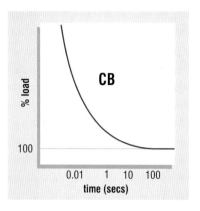

For normal (non-surging) electrical loads, fuses and CBs are to a large extent interchangeable.

All too often there is a misguided perception about the function of the main circuit fuses or CBs which needs to be clear in the minds of all boat owners. Protection devices are deployed to protect the wiring and switches, not to safeguard equipment. Any trip or fuse should be rated at a higher current than the total current expected to be drawn under steady no-fault conditions Then, most important, the wiring which the fuse protects should be capable of carrying more than the rated current of the fuse or trip without dangerous overheating. The protection of the equipment must be by a separate fuse sited locally or built into the device itself by the manufacturer.

RUPTURE CAPACITY

An ordinary cranking battery is quite able to put out 2000 Amps into a dead short, and a powerful battery bank can provide significantly more. When a short occurs, the current is limited only by the resistance of the wiring involved, and a current of hundreds of Amps (in some circumstances, thousands) can quickly build up, in a fraction of a second, before it is interrupted by a breaker or a fuse. Breaking a current of this magnitude can be quite a brutal process, involving the formation, and then the breaking, of an arc. With a fuse, there can be a miniature explosion of molten and evaporated metal. With a circuit breaker, there can be a danger of the arc causing erosion of the breaking contacts, or worse, welding together those contacts.

To handle this hazard, special HRC (high rupture capacity) fuses are available, which may embed the fuse wire in a sandy powder. For some breakers, the rupturing capacity is specified; for others, the data sheets are silent. A device with an ability to break high currents may be used as the first line of defence; the first protection downstream of the battery.

Circuit Breakers

These are more expensive but a long-term investment. And, because they also double as switches, some of the cost is clawed back because you would need separate switches if you'd chosen fuses. Apart from predictability and reliability, another advantage is that a tripped CB can be identified in the dark by feel. Of the five or so general types of CB, just two are commonly used on boats, namely thermal or magnetic CBs.

Thermal CB

This uses a bimetallic strip heated by the current passing through it. As it heats it gradually deflects the strip until it unlatches the mechanism holding the contacts together. Because it takes time for the heat to take effect, this forms a convenient delay to accommodate any current peaks. And similarly, once tripped, the bimetallic strip must first cool down before the device can be reset - and in so doing, also allowing time for the circuit and appliance to do likewise. This is called a trip free mechanism and means that the switch and its contacts cannot be held against a fault condition. Without this mechanism the CB could be immediately reset, thus enabling the current and the heat to accumulate to cause an electrical fire.

Thermal CBs are also self-compensating when it comes to ambient temperatures. Because the strip automatically tracks the external heating effects on the cable, it will trip without delay once the heat threshold is crossed. For example, the electrics of a yacht in the tropics will be at a higher temperature than the electrics of a yacht in UK waters. Even without any current flowing in the wiring the temperature of the cabling in the tropics could already be as high as 40-45°C - fairly warm! Just by the influence of the surrounding ambient temperature alone, the bimetallic strip will be partially on the way to tripping the CB. With the addition of a current flowing through the sensing element, more heat will be induced and the strip will move that bit further towards tripping, thus requiring the CB to operate

within a very tight margin and be capable of tripping that much sooner. It follows that a yacht operating in cool conditions will have a greater margin of tolerance.

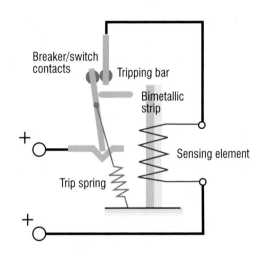

Breaker/switch contacts — Tripping bar — Bimetallic strip — Sensing element — Trip spring

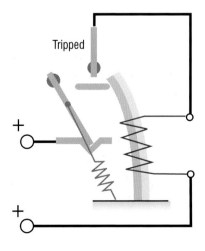

Tripped

Magnetic CB

This uses a solenoid to work against a spring as its operating mechanism, and as such makes its operation very fast. These CBs can be reset immediately after the instance of tripping and if the fault is still present then the CB will trip again immediately. This will conveniently make sure that there's simply no time for a lot of heat to be dissipated should the CB be repeatedly reset after tripping.

Since these CBs are so fast and prone to spurious tripping through shock, vibration and the slightest overload, they are not usually preferred. Some magnetic CBs incorporate a thermal delay mechanism to dampen down the sensitivity for the smaller overloads, leaving the magnetic mechanism to provide immediate tripping for large overloads.

Marine life asks a lot from CBs. Indeed they may sit there, in a closed condition, quietly passing current for years. Then, quite suddenly upon an overload or short, they're required to work with 100% reliability. Whatever the type of CB, they should be exercised regularly by switching them on and off. This will keep the pivots free to move and any corrosion from the contacts.

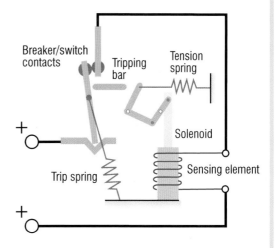

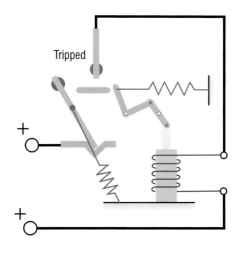

The Wiring Loom

LED: Light Emitting Diode
Like a normal semi-conductor diode but this time doped with different impurities whereby the free electrons flowing across the diode release extra bands of energy which is radiated in the form of visible light. The colour of light depends upon the energy bands released. LEDs are usually combined with a resistor to limit the current otherwise they will blow.

To save time and cost, most production boats are fitted with specially prepared wiring looms to serve the various electrical appliances dotted around the boat. The batteries, distribution panel and the main body of the loom usually lie on one side of the vessel, with individual wiring reaching out to supply current to wherever it's needed.

Although convenient for the builder, looms can be a problem for any boat owner wanting to add appliances at a later date. If the boatbuilder has been thoughtful, he may have installed a loom which contains spare cables – up to 20% extra being typical. If not, the owner faces the risky business of tapping into the existing distribution, not knowing whether or not this might overload the wiring, particularly the DC return linkage which may be serving a number of different circuits. If in doubt, a separate return should be run either directly to the battery negative or, if there is one, to the mid point of the horseshoe shunt.

Unless specifically provided for in the original loom, power hungry devices, such as electric anchor winches and bow thrusters, will almost certainly need their own supplies.

A distribution panel dropped down to reveal the bus bar links, junction terminals and the cable looms. The bundles of red and black cable looms will snake away to serve their respective DC appliances.

Distribution Panel

This illustration shows a full panel board serving the AC services at the top and below that, the 12V DC distribution network. All the CBs on the DC board are clearly labelled and its quite usual to group or theme the CBs into logical sections like: Nav Lights, Services and Electronics, etc. Note that one or two CBs are left aside as spare. A futher group of CBs would be sectioned off if you had a maintained supply which is discussed under 'Don't switch off' (page 73). Here, the CBs are 'reset type only' where the CB can be manually closed but only opened through tripping. Each CB will have its own indicator light to display at a glance which circuit is on line. Also, notice that the LED indicator lamps for the Nav light group have been repositioned to the yacht mimic,

providing a very graphic display of the navigation light status. Instrumentation is usually brought to the panel board as well, and at the very least one would display the network current and the battery voltage. Alternatively, sophisticated battery monitors can be brought to the panel board which can process and display an array of information about the condition of the batteries, including their voltage and load current and even the number of hours remaining at the current rate of consumption. Some panel boards will extend their monitoring facilities to include fuel and fresh water level gauges

Mast Wiring

When masts are delivered, they usually come pre-wired, so the manufacturer or owner simply has to trust that it's all been done properly. Let's look at some of the wiring features you might find.

Extra weight aloft is never a good idea, and some savings – including those of cost – can be made by sharing the negative returns. This trick is often used advantageously with the tricolour and anchor lights, and the steaming and deck lights, particularly when those combinations are built into the same unit. In both cases, the individual lights are powered by switched (and fused) positive feeds, with the return going back down the common negative. For simplicity's sake, 3-core flexible cable is usually used. Incidentally, on no account should the mast itself be used as the return, as this would introduce a very considerable risk of short circuits and could also cause serious electrolytic corrosion to the aluminium extrusion.

With the aid of complex switching and the cunning use of diodes, there are ingenious arrangements that can economise on the mast wiring even further, but the benefits of going to so much trouble are doubtful. However, diodes can be useful in certain circumstances, one of which is shown here. Note the junction box in the port quarter of the drawing on these pages in the positive feed to the compass light. Tracing the wiring back you'll see that

when darkness falls and either the nav lights or the tricolour are switched on, power will also be supplied to the compass light. The diodes prevent the nav light's current feeding back into the tricolour's circuit and vice versa.

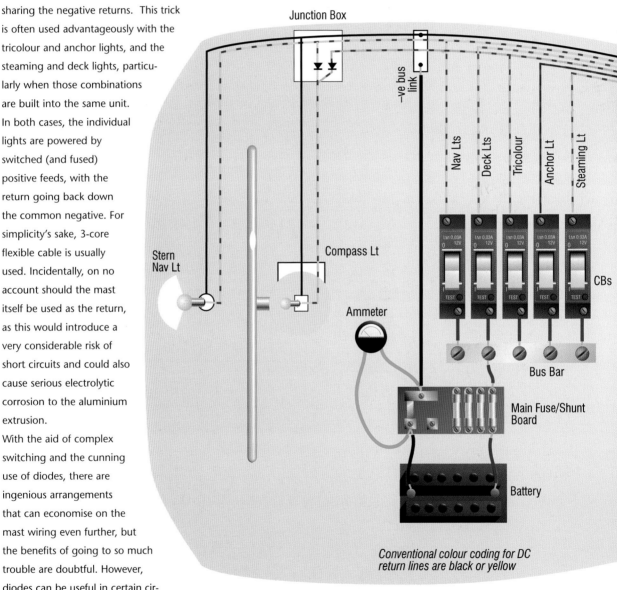

Conventional colour coding for DC return lines are black or yellow

Mast Foot Glands & Deck Plugs

Keeping water out of your mast wiring is always a headache, and the most vulnerable points are where there are connections. Masts must occasionally be taken down, usually to be racked off ashore or on deck, so it's impracticable for the wiring to be permanently integrated with the distribution circuits that supply them. To get over this problem, it's usual to have some sort of connection at deck level, either above or below. There are strengths and weaknesses to either of these approaches. If connecting or disconnecting the wiring is to be done above decks, you obviously need some sort of waterproof plug – and it had better be a good one, if you don't want to see it flooded. Quality really counts with these. It also helps to apply some petroleum jelly (Vaseline) to the seals and threads, but never to the actual contacts because the jelly will partially insulate them and could cause a drop in current and dim the nav lights. Always keep your contacts dry and clean. Although perhaps seeming the more awkward arrangement, having the connections below gives them much better protection. The cables usually pass through the deck via some sort of deck gland which forms a seal around the cable. Obviously, figure-of-eight or oval cables are more difficult to seal, so it's much better to use round sectioned cables.

Once below decks, the various wires will terminate at a junction box, placed as close as is practicable to the mast foot (assuming it's deck stepped). Just in case the gland might leak a little, it helps to form a small loop in each cable before it enters the junction box – that way, any water droplets will drip off the bottom of the loop before they reach those all important connections. Again, buy the best quality gland you can find.

Junction Box

Junction Boxes

Deck Lt

Port Nav Lt

Stbd Nav Lt

Steaming Lt

Deck Lt

Anchor Lt

Tricolour Lt

Mast foot/deck glands

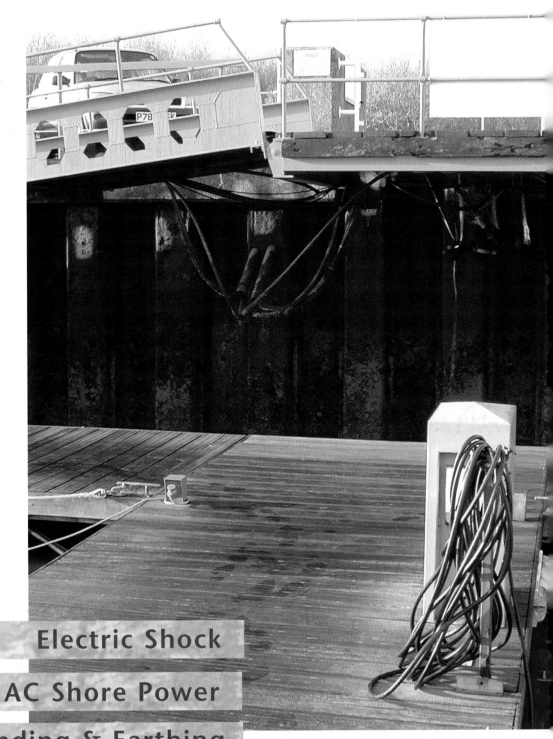

Electric Shock

AC Shore Power

Grounding & Earthing

RCDs

Bonding

AC Safety

In addition to their 12V DC system, more and more new boats entering the market are going to have AC, or the ability to receive AC, as a standard fitment. Bringing 240V AC aboard enables owners to power the sort of facilities – heating, refrigeration, microwave, TV, etc – that they are accustomed to enjoying at home. Some second-hand boat owners may aspire to having this facility themselves, and there are plenty of electrical firms out in the market place eager to provide custom installations. The simplest AC supply may simply be an extension lead plugged into a marina socket at one end, trailed through the companionway, and running, say, a portable fan heater or battery charger at the other. Others are permanent installations of considerable sophistication. Whatever the choice, the incorporation of AC power on a boat brings with it new problems and a slightly different attitude with regard to safety. The protection of the cabling and its associated switchgear in an AC circuit, adopts the same principles and philosophy as in a DC system, but since the voltages in AC are elevated to 240V there needs to be greater emphasis on the protection and preservation of human life. This factor is emphasised even further with the knowledge that these high voltages are being carried within a waterborne environment, making the threat to human life that much greater. It is therefore essential to have a thorough knowledge of AC safety and the dangers that this electrical medium can bring.

7

Shocking Stuff!

When dealing with mains electricity we all know that it can kill, and the risk is elevated when we use AC shore power in a water borne environment to provide power to boats. The main bulk of the human body is built up of a composition similar to salt water, and functions through small electrical nerve impulses sourced from the brain. Understandably, then, the human body will have a violent reaction to a large electrical current passing through it. The behaviour of the human body when under an electric shock is involuntary, but can to a certain extent be predictable.

DC shocks under certain conditions can be lethal; generally speaking the danger of shock is not nearly so great as with corresponding AC supplies. A large number of marinas are now equipped with 440/220V AC, and a shock from these voltages may easily be fatal. One of the effects of AC, apart from other considerations, is that it causes a tightening of the grip, which often means that the unfortunate victim cannot let go. Electricians have fairly mixed views when it comes to human reactions under DC or AC shocks, but reference can be drawn from the Institute of Electrical Engineers, ref publication: PD 6519, Part 1: 1995 Section 1 *Guide to Effects of Current on Human Beings and Livestock.*

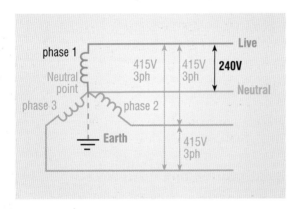

The nearest neutral point to an AC distribution system is known as the source or origin of power.

What is little appreciated is that it's not so much the voltage but the current that actually kills. A current as little as 200mA from an AC voltage of 60V is more than enough to be fatal, and obviously a higher voltage would increase the current flow. The most lethal shock is where the path of the current passes from hand to hand through the chest and, if sustained long enough, the rhythmic cardiac beats of the heart can be interrupted or, worse, stopped.

From the diagram (far left) we see that AC power is sourced as 415V – 3 phase, and most house-holds/boats, etc. receive their AC as 240V single phase. The reduction to 240V is achieved by taking the power from one of the three spurs in the 415V system, each spur or phase being 240V. The distribution cable from this phase is considered the live wire. The centre or meeting point of the three spurs is the neutral point, and the distribution cable connected here is considered the neutral wire. The neutral point is also earthed, where a cable is led to a large metal plate or pen buried underground. The nearest neutral point to an AC distribution system is known as the source or origin of power. Through our home DIY, we are aware of the conventional colour coding of these cables, where the live wire is brown, the neutral wire is blue and the earth lead is yellow and green.

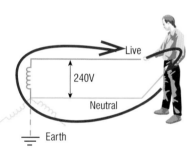

We can only receive a shock if we become part of an electrical circuit in one of two ways.

Firstly, we can place ourselves bodily between a live and neutral wire, making the circuit as shown.

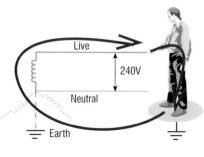

Secondly, in contact with the live wire, we can form a path to earth, making the circuit as shown. In the latter circumstance, were we to jump up in the air whilst still touching the live wire, we would momentarily be relieved of the shock since the circuit would briefly have been broken.

This example explains why birds sitting on suspended 40,000V high tension cables do not receive a shock. Incidentally, the wearing of wellies or standing on a dry fibreglass deck would make you relatively safe, but only the foolish would rely on that.

It's not always necessary to touch a live wire to receive a shock. Most electrical appliances are encapsulated in a metal case and the live and neutral wires are intended to be isolated from each other and from the metal case. A live wire which is displaced from its terminal fastening would leave the exposed conductor susceptible to touching the metal case, thus rendering the case live. Alternatively, a similar result can occur when the electrical insulation between the live wire and the case

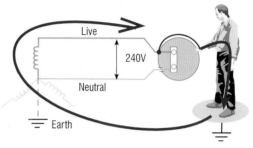

breaks down. The latter situation has a nasty edge, since the operation of the appliance remains unaffected, but a person innocently coming into contact with the case will receive a full earth-ground shock. These 'live to ground' shocks, which are created by faulty wiring and where the appliance or the metal casing around the appliance becomes live, are probably the most common.

Once appliances are earthed, however, then should any faults occur, the live current must ultimately find its way back to the source of power so that we can then encourage a short circuit to develop, making the circuit as shown. If the appliance is fused on the live side and if the earth return circuit has a low enough resistance, then the short circuit current will blow the fuse, or trip

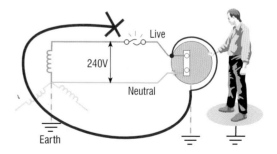

the CB isolating the appliance thus rendering it safe. It makes sense, therefore, to have the source of power as close to the load as possible. If, however, the appliance wasn't properly earthed, our victim would still receive a shock since he's now forming part of a parallel circuit.

Marinas

A pontoon pedestal

The great majority of AC power coming on board a boat is going to come from a marina and most marinas in the UK and Europe will be AC receptive. It makes sense, therefore, to have some knowledge of a marina set-up and the safety devices that are installed for the benefit and protection of you and your boat.

Below is a typical marina set-up. The regional electricity company supplies the marina company with 415V 3-phase AC power from a nearby delta/star transformer with the star neutral point earthed. The marina company takes over authority at the meter/isolator and supplies electricity (still 415V 3-phase AC) to a main shore distribution pillar and then over the water to a pontoon distribution pillar.

From here the three phases are split and distributed to the individual berths as 240V single phase, terminating at the power pedestals or 'Daleks' into which the boat's power leads are plugged. Typically each plug is metered, and here the responsibility of the marina ends. The rest is up to the boat owner.

From the supply pedestals back to the marina isolator there's a discriminatory sequence of breakers and protection devices as well as an earthing system. The pedestal will have a type 'C' MCB set to trip at 30A, and an RCD which will trip if a differential

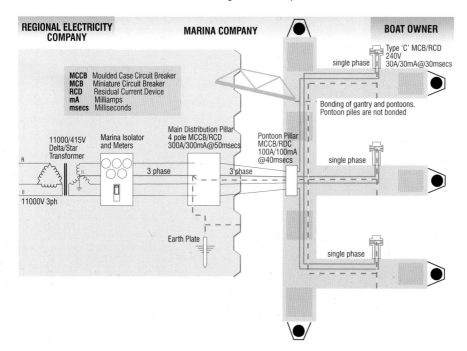

current of 30mA is sustained for more than 30msecs. The pontoon distribution pillar will have an MCCB set to trip at 100A with an RCD set to 350mA at 100msecs, and the main shore distribution pillar will have a 4-pole MCCB set to trip at 100A with an RCD set at 1A at 0.5secs. All the AC cabling will have an earth lead and all the piles, floats and gantries will be bonded to it. The main shore distribution pillar will have an earth plate buried in the ground and will be the nearest earth point to the waterborne marina network; it is therefore the most important earth point of the marina system. Should the shoreline be faced with steel corrugated piles then these will be bonded to the earth plate. It should be appreciated that a marina is a place where the risk of electrical leakage is immense – meaning that a marina could be a fairly corrosive environment for any boat.

RCD

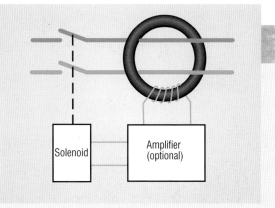

There are two sides to an electrical circuit; for in DC there's the +ve side and the -ve side and for AC circuits there's the live and neutral. With regard to AC, if 10A goes down the live wire then we would expect 10A to come back along the neutral wire; if not, we have what's called a leakage current. Somewhere in the circuit the insulation must have failed and as a consequence something may be dangerously raised to live potential. Ideally we would hope that the leakage, wherever it may be, would travel to ground via the earth wire and blow the fuse in the live wire. However this may not happen if the ground/earth resistance is high, so the potential shock could still kill. What's needed is a faster, safer, and more sensitive piece of equipment like a Residual Current Device (RCD).

In the RCD both the live and neutral wires pass through a circular iron ring. The incoming current from the live wire induces a magnetic field in the ring in one direction and the equal and opposite current from the neutral wire induces a magnetic field in the opposite

direction. The net magnetic field detected at the sensing coil is therefore zero. Any difference in the two currents (as a result of a leakage somewhere) produces a net magnetic field which induces a current in the sensing coil wound around the ring. Sometimes to enhance the sensitivity of these devices, the sensing signals are amplified to activate a solenoid tripping mechanism, disconnecting both the live and neutral sides. To demonstrate their sensitivity, the detected differences in current are rated in terms of milliamps and the speed of activation is rated in terms of fractions of a second.

See Appendix F for Leakage current

House or Boat?

Bringing AC shore power on board a boat in a marine environment is dangerous enough and it becomes more dangerous when people make the mistake of applying shoreside DIY practices to the world afloat. Conflict or confusion occurs between earthing the appliance, earthing the boat and earthing to the shore network. Part knowledge of shore practices can be

DIY Roll Cord Extension

Our first owner has a simple roll cord extension cable which he plugs into the marina and brings the socket end inside the boat for general use. And what can be wrong with that? After all it's a three core cable whose earth leads directly to the earth on the supply and, anyway, all of his appliances are fused at their plugs. Well let's assume he's the unlucky owner of a faulty appliance and the casing has become live. Once plugged in, he's then immediately reliant on the efficiency of each link in the earthing chain to carry that current ashore and bring it back again to blow the fuse in the appliance's plug. Should there be enough resistance in the earth chain

and its connections, say through rust or dirt, then the appliance may be considerably above ground potential. Now the owner may find that the alternative circuit shown offers less resistance to the above route and will find himself forming part of a parallel circuit to the earth cable system. Because of this additional parallel load placed on the circuit, the current flow through his body is still insufficient to blow the fuse. Current may flow through him, sufficient to cause a lethal shock. We should also remember that the fused plug isn't there to protect the owner, anyway, because it's designed to stand a much higher blowing current than he can! Its purpose is to protect the circuitry from an overload and risk of fire.

Theoretical Solution: All AC appliances should be earthed and bonded to the boat's ground, thereby bringing all the AC appliances and metal skin fittings to earth/ground potential.

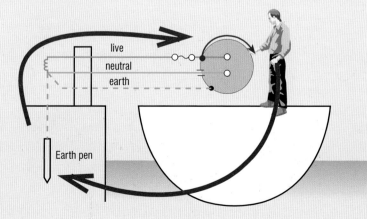

live
neutral
earth

Earth pen

equally dangerous when let loose on board, for here both the issue of earthing and the location of the neutral point can play havoc in the overall safety of the vessel. The two most common scenarios are featured below and from these we can learn a great deal from the mistakes and the dangers that lie within.

Earthing the Neutral

The second owner has a bit more electrical knowledge than the first and so sets up a more permanent installation. Mindful of the problems faced in the previous example, he draws on his knowledge and the theory of shore practices in that the neutral is earthed. He connects the boat's earth wiring to the neutral wire at the onboard AC panel, reasoning that this will give him the additional safeguard of an earth return through the neutral wire should the proper earth wire develop any problems. Assuming the same faulty appliance as above, we can see in the diagram that the system has created three parallel return paths. Now, should the owner's concern over the earth cable become apparent, then nearly all the earth return current would choose by far the easiest route - straight down to the water to the boat's ground. Imagine the consequences of a swimmer in the vicinity of the boat! It's worth noting that in this example the owner in

effect has connected the neutral to the sea. With an RCD connected to the system, it would immediately trip and render the boat and the swimmer safe. The owner's main misconception here is over the theory of earthing the neutral. His knowledge and understanding of shore practices in earthing the neutral is partly correct for the neutral is earthed - but only at the source of power (which is usually the local transformer serving the street) and not at the distribution board.

Worse - should the boat be connected up with reverse polarity - a common occurrence - then the live side will be shorted directly to ground and the hull fittings via the neutral connection, and very high current will flow.

Theoretical Solution: The earth and neutral wires must not be connected on the boat, but at the proper shore power source.

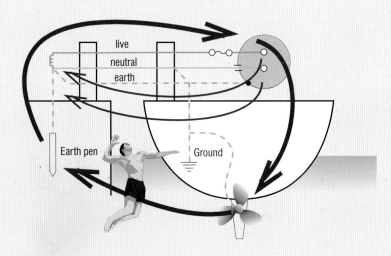

live
neutral
earth

Earth pen

Ground

To Ground or not to Ground?

The debate rages on the important and potentially dangerous matter of whether the AC earthing should be common with the DC ground (or negative bus link).

The position adopted by the American Boat and Yacht Council (ABYC) and the British Marine Electronics Association (BMEA) – and taken up by the European Recreational Craft Directive (RCD) – is that any boat connected to shore power should have its major metal fittings and any large conductive masses in contact with the water brought to earth potential to avoid any risk of shock (left).

Then, on the other hand, the US Power Squadron and others take the view that AC and DC are different in both character and voltage and therefore should be kept quite separate (right). Their concern is that should, say, poor connections or faulty and degraded cables create a high resistance in the AC earth, then the DC

ground could provide an alternative path through the water, with possible risk to people swimming nearby (see P.89). Also there's the risk that if the DC system were improperly bonded to the AC earth system, then an AC insulation failure could bring the supposedly 'safe' DC system up to AC voltage with obvious dangers to the crew.

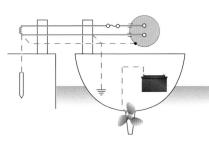

So far the issues focus mainly on the safety of human life. But as soon as owners find out that these safety factors can jeopardise the welfare of their boat then the main impetus of the debate becomes a straight fight between the welfare of the boat and the safety of human life. Surprising, or perhaps not so surprisingly, some owners are prepared to put the welfare of their boat first. It helps, therefore, to understand the problem that lies behind this debate and look at the various ways of solving the issues.

Earth Loop

No personal dangers here but a bitter irony for conscientious boat owners whose electrical installations are otherwise beyond reproach. Below is a pair of yachts, each connected properly to the shore supply, and thus also connected to each other because all boats in the marina share the same common earth lead. Also, both yachts are conforming to EU practices, where the AC earth is bonded to the DC ground. Unfortunately there's now a galvanic relationship between the two vessels and the direction of the galvanic circuit is of course dependent upon which of the two boats is more anodic to the other, the anodic one being the one to suffer serious corrosion. The rate at

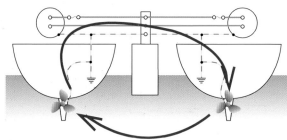

which this corrosion takes place can be concerning as marinas are becoming quite corrosively hostile environments. With this in mind and the prospect of some expensive annual maintenance bills, some owners are prepared to sever this corrosive earth loop, with dangerous consequences. On the whole, these owners are prepared to **BREAK THE BOAT'S EARTH** connection or **FLOAT THE BOAT'S AC SYSTEM** by breaking its connection to the DC ground.

Ideally, what the owner needs is a device which will allow a full AC earth fault current to flow to ground but will not allow small galvanic DC currents to leave or enter the boat. Two choices are available to boat owners - the **GALVANIC ISOLATOR** or the **ISOLATION TRANSFORMER**.

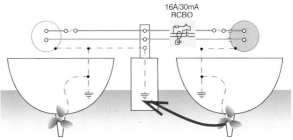

Broken Earth

To break the earth loop, the earth connection is not connected to the boat. The AC feeds the distribution panel via an RCD and all the AC appliances are earthed and bonded to the DC ground. Primary protection is placed upon the RCD and reliance is placed on the DC ground and hull fittings to carry any earth leakage current. There could be sufficient resistance or other discontinuity in the ground system to allow any leakage current to flow and thus possibly prevent the RCD from tripping. This arrangement is technically sound but has its risks making it, for safety reasons, not recommended.

Despite the earth loop being broken, leakage currents from electrical equipment to ground could exist, prompting localised electrolytic action - so the corrosion problem could remain.

Floating AC System

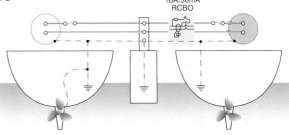

The shoreside earth connections are retained but are not connected to the boat's ground - sometimes referred to as a 'floating AC system'. Reliance is placed on the shore earth cabling to carry all the fault currents which does pose a risk, for if there were any lack of continuity in the earth line due to faulty or degraded earth terminals this could leave the earth fault current with no place safely to go, posing a very serious shock hazard. Those that select this approach and have onboard generators or inverters could find themselves back to square one, for the frames of these additional power sources are sometimes accidentally or deliberately by design connected to the boat's ground, thus closing the earth loop again.

Galvanic Isolator

In simplistic terms this consists of a pair of opposing diodes in parallel. Any alternating current is assured one of the two routes through the diode circuit. As for any DC current, the reverse biased diode will block any DC current outright, and the forward biased diode will block any DC current up to 0.7V, (remember a diode in a forward direction has a small voltage drop, typically about 0.7V to allow current to pass in the forward direction). If a galvanic DC blockage of 0.7V is not enough, then simply double up the diodes in series to total 1.4V.

Isolation Transformer

This is a 1:1 transformer which neither steps up nor steps down the voltage, and therefore the input and output voltage remains at 240V AC. Because the current is passed on by induction, the input and the output lines aren't actually connected to each other at all. The earth leads are also isolated, since the shore earth terminates at the transformer shield core. The boat's earth network is brought to the neutral point of the transformer output so the boat is isolated from any leakage or stray currents.

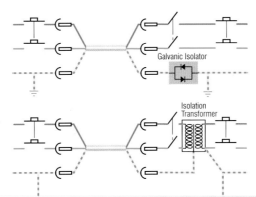

Galvanic Isolator

Isolation Transformer

See Appendix F for Leakage Current

Bond! Dr.Yes or Dr.No

However perfect the system and however meticulous its installation, as soon as we have shore power corrosion problems arise. This is largely due to stray currents originating either from within the boat or from shoreside devices or neighbouring boats. In all cases leakages from current-carrying cables will find a path to ground or earth through bilge water, damp areas of the boat and sea water. All of the boat's internal and immersed metal fittings will be at different potentials and, especially in marinas, the water outside will vary in potential from place to place.

We can think of the ground or earth as being the potential of the water in which the boat floats, and the purpose of bonding is to tie all of the boat's metal components together to try to force their various voltage potentials to be as uniform as possible. The bonding cable should be a very low resistance conductor, entirely separate from both AC earthing or DC grounding – although they can, by design, all terminate at a common point. If the bonding network is anchored to DC ground and AC earth (as on P.90) it will ensure that a fault current in any component will eventually flow by the appropriate earth cabling, and not by a parallel path through the water(see P.89). Bonding therefore extends the earthing system, but ironically, bonding will create a galvanic circuit and can actively encourage stray currents and accelerate the corrosion problem. For this reason, a bonding system will have to incorporate an anodic protection system to protect all components included in the network. But nothing is ever that simple. Let's explore how the various dilemmas present themselves, and how problems can be minimised by balance and compromise.

See Appendix F for Leakage /Stray Current

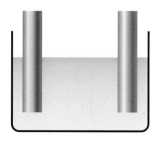

If you have two dissimilar metals (or alloys) in a tub of sea water and they are electrically unconnected, then the worst that can happen to them is a natural oxidation or rusting. No current will exist between them but there will be a potential difference between the two dissimilar metals. This could be the situation, say, with a pair of isolated skin fittings.

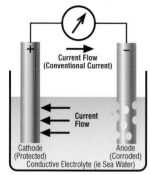

Once the two metals or alloys are linked by a conductor, the circuit is complete and we have a simple galvanic cell. A current will be produced at the expense of one of the metals and this process is known as galvanic corrosion. A practical example could be where an owner bonds two underwater fittings together for reasons of AC safety, and suffers corrosion to one of them.

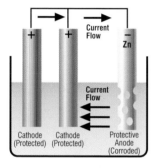

By introducing a third metal – commonly zinc – we can save the situation. Zinc was chosen because, like all ignoble metals, it relishes being part of a galvanic cell, and through its own greed will take most of the galvanic current and will physically sacrifice itself in its eagerness. This is the principle of the sacrificial anodes bolted to the underside of most hulls.

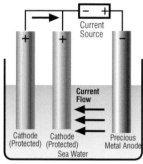

Alternatively, we could do the opposite and replace our zinc with a metal with less enthusiasm for self-sacrifice – a chrome alloy perhaps (see Galvanic Table P.19). The current in our tub normally flows from left to right, causing one of the two metals on the left to corrode. But if we introduce a controlled current into the metal on the extreme right, we can counter the natural galvanic flow by creating an artificially impressed cell at a potential just above the galvanic voltage (say 0.2V). Galvanic corrosion is now halted by what's called 'impressed current protection' a common provision on commercial ships.

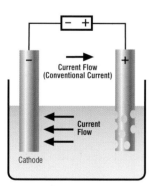

Attractive though the last method might seem, there's a sinister possibility. As we know, when two similar metals are immersed in an electrolyte and bonded together, there will be no potential difference and consequently no current flow or corrosion. But if we then place a current source between the metals we'll impress an electrolytic circuit and the current-receiving metal will corrode. This situation can arise accidentally on boats, where a poorly insulated conductor can release stray currents, possibly of relatively massive voltage – 12V or more, rather than a few tenths.

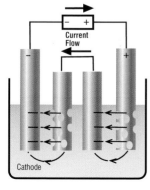

If two innocent bonded objects (of similar metal) are caught in the path of a stray current through the sea water, they'll become an impressed electrolytic cell where one of the objects will corrode.

The top left boat is floating in an external voltage and current field, as might be found in a marina. If the electrical resistance in the green bonding cable is less than that of the water, the seacock will pick up the current, pass it along the bonding cable and out through the propeller, and the propeller will corrode. A sacrificial shaft anode mounted on the prop shaft (see opposite) would divert the current away from the propeller.

Electrolytic or galvanic corrosion? There is often some confusion when identifying electrolytic action (electrolysis) and galvanic action.

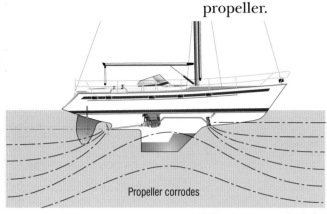
Propeller corrodes

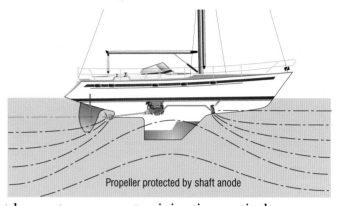
Propeller protected by shaft anode

GALVANIC ACTION
This is the result of an electric current developed by the immersion of two dissimilar metals in a conducting medium. The value of the current is not influenced by any external situation other than previously stated.

The bottom left boat has a stray current originating entirely from within. Both the bilge pump and seacock are submerged in bilge water. The unbonded pump develops a short between its positive lead (red) and its casing. A current then takes the path through the bilge water to the seacock, out into the seawater, back to the propeller and shaft, and through the engine block to ground. In these circumstances the seacock would corrode, but if the pump and the seacock were bonded (as shown bottom right), the stray current would have chosen the less resistant path along the bonding cable (green) and there would be no corrosion.

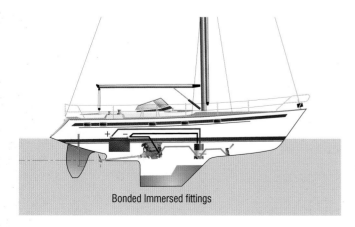
Bonded Immersed fittings

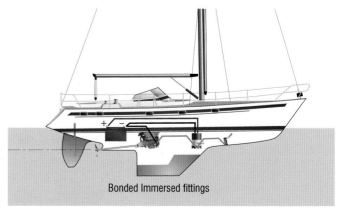
Bonded Immersed fittings

See Appendix for Leakage /Stray Current

From these examples we can generally conclude that bonding immersed fittings prevents corrosion from stray currents from inside the boat but still leaves them prey to currents from outside the boat's hull. This is a dilemma for which there are two solutions: we can either 'bond and protect', where all immersed metal fittings are bonded together and employ a sacrificial anode to defend the immersed fittings (top boat), or we can 'unbond and isolate' (bottom boat), where the shaft is isolated by a flexible coupling and the propeller protected by a zinc shaft anode and all other immersed metal fittings are unbonded, thus ensuring that neither galvanic nor stray currents can flow between them. All internal fittings are bonded together so that, for boats with AC supplies, these fittings are placed at ground potential.

To the left of the P-bracket is a typical prop shaft anode. Many, like this this one, are streamlined. The gadget between the propeller and the P-bracket is a rope cutter.

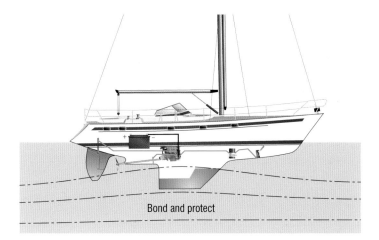

Bond and protect

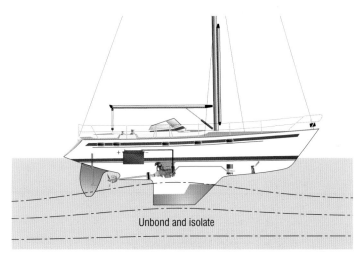

Unbond and isolate

ELECTROLYTIC ACTION
This is the result of a current from an external source, e.g. a boat's battery system. Effectively this means that two similar metals can form the anode and cathode. The rate of electrolytic action depends on the value of the stray current, which can be anything from a trickle due to dampness, or at the other extreme, a short circuit where there are no inherent limitations, as in the case of galvanic action.

Battery Chargers/Inverters

AC Polarity

RCBOs

Transformer & Galvanic Isolators

AC Distribution

Having an AC supply delivered to your marina berth is proving far too tempting an opportunity for more and more yacht owners. Not only can they enjoy many of the same amenities they have at home but, so long as they remain alongside, their batteries should always be fully charged.

In the last chapter we looked at some of the potential dangers. Now we turn to the various onboard systems and the devices that help make them safe.

Onboard AC distribution can be divided into one of three categories:

• The owner with a portable cable feeding a single appliance only;

• The owner with a permanent system, served through a distribution board and feeding just a few appliances;

• And, a full blown AC installation which includes sources of AC other than shore power.

Let's look at them one by one.

Portable Power

Below is an arrangement which the majority of boat owners will be familiar with. An extension cord is plugged into the pontoon supply and is used to power the battery charger. From then on the boat essentially runs on its 12V system, with the battery levels maintained by the charger.

The most worrying aspect of this set-up is that there's no polarity warning light (see Poles Apart P.104). But even if the polarity were correct, the degree of protection is meagre, relying as it does on fused plugs (and sockets, sometimes) and the earth cable to shore.

The prudent owner will also fit a residual current device (RCD) as described on P. 87, either a spur mounted RCD device, wired directly to the appliance, or an RCD adaptor plug between your ordinary plug and socket. Indeed, there's a powerful argument in favour of making an RCD the absolute minimum protection you should consider. They're not particularly expensive – typically £12 or so – yet offer a considerable improvement in safety.

The connection with the marina pedestal can vary from place to place, but whatever arrangement is used it must be watertight and lockable to ensure that it is both electrically and mechanically sound.

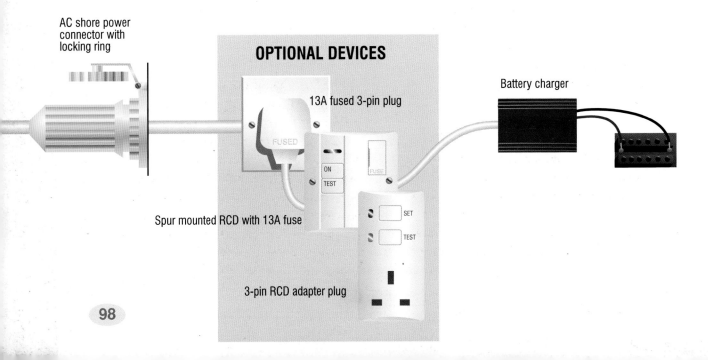

AC shore power connector with locking ring

OPTIONAL DEVICES

13A fused 3-pin plug

FUSED

Battery charger

ON
TEST

FUSE

Spur mounted RCD with 13A fuse

SET

TEST

3-pin RCD adapter plug

2 Pole Breakers

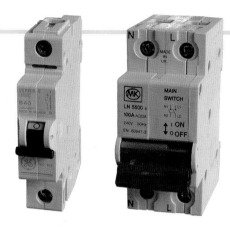

Whatever type of circuit breaker is used, it is essential to make sure that it is a two pole variety. Household CBs are usually single pole breakers (left), meaning that as a result of a fault condition only the live circuit is broken.

If such a breaker were adopted on board a boat with a reverse polarity situation, then the breaker would operate on the neutral side of the circuit. Although still available to work effectively as a switch, the appliance would unfortunately, remain live and potentially hazardous. Clearly two pole CBs or 'phase and neutral CBs' (right) cut both the live and neutral lines simul-

taneously, rendering the circuit safe irrespective of polarity. The phase or live side of the CB usually tracks the current rating, but upon tripping it will isolate both the phase and neutral lines.

Battery Chargers

Mains chargers work by stepping down the AC to 12V through a transformer and then rectifying the output through a network of diodes to produce 12V DC. You'll also get some form of voltage control, the sophistication of which is largely dependent on how much you're prepared to pay.
(see Smart Charging Controllers P.54 & 55)

The simplest units are the low cost types available in car accessory shops. These chargers are usually unregulated and should only be used as a temporary expedient to give a charge to a flat battery. Although certainly capable of charging a smallish battery, they simply haven't enough grunt to make any real impression on a large bank of them. Also, with their loose leads and crocodile clips, it's very easy to spark them or get the polarities wrong.

But, if you're the type that draws large amounts of power from your batteries, then an arrangement as shown on P. 101 may be the way to go, with a larger output charger wired permanently into the system. Such chargers typically have outputs between 10A and 100A and may have two outputs, enabling the charger to support more than one battery or bank of batteries. The charger's size should be matched to the capacity of the batteries they serve and, of course, your demand. A good rule of thumb is to have the output of the charger rated at 20% of the battery's rated capacity - i.e. for a 100Ah battery, a charging current of 20A is ideal. Higher rate chargers can be used with 3-stage regulators which will of course prevent the batteries from

being overcharged, but one has to weigh this against the extra cost.

Some chargers are specifically designed or have the adjustments to suit the type of batteries used, i.e. lead-acid or gel type. Additional adjustments could be in the form of temperature compensation, since low ambient temperatures require higher charging voltages. Good chargers will provide this as an automatic feature.

An equalising charge switch can be found on some mains chargers, and this is used every few months or so to condition the batteries (see Equalising P.67) and clean the sulphate from the battery plates.

Fixed systems

For a fixed AC system, only multi-strand 3-core cable should be used – not household type single strand cable. Over time copper does 'work harden' through the external influences of temperature, tension and vibration and eventually becomes brittle. This would leave a single strand cable vulnerable to snapping or breaking free, say, from a terminal. Having an exposed and uninsulated conductor core at 240V has obvious hazards; the installation of the more ductile multistrand cables will alleviate the problem.

Ensuring the safety of a fixed system calls for both earth leakage and over-current protection – a residual current device (RCD) and miniature circuit breaker (MCB) respectively. There's much debate over which should precede the other, but many would say that it doesn't make much difference.

Today, most people would install a residual current breaker overload (RCBO) device which combines an RCD and MCB in a single unit. A typical RCBO would trip on an overload of 16A and a residual current difference of 30mA in 30msecs.

The Recreational Craft Directive dictates that the MCB/RCD must be within 0.5m of the shore cable, but it's quite usual to have such devices mounted on the boat's AC distribution board – not exactly pleasing to look at but certainly convenient. The panel board will also usually include instrumentation and a polarity check light.

In many ways, fixed AC systems are similar to a 12V DC distribution. Each live wire (brown) from the distribution board has its own direct lead to its respective appliance, and the neutral wire (blue) returns the current to a common neutral terminal. Even the CB indicator light adopts the same principles as its DC counterpart. Where AC distribution differs from DC, however, is that

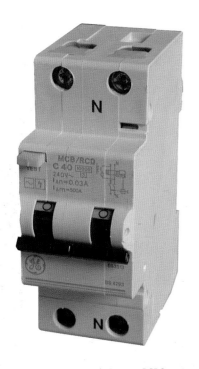

This is a MCB/RCD or a RCBO. It's a type 'C' breaker, designed to trip both phase and neutral upon an over-current rating of 40A and a residual current rating of 0.03A or 30mA.

The yellow button is the manual test device - press it, and it should trip. This test should be carried out at least once a year and noted in the boat's log.

all cables leading to AC sockets must have a third cable to carry the earth. The voltmeter and the AC power light naturally straddle the bus bar and the neutral link, but the ammeter receives its current signals from a toroidal current transformer such that the full load current does not have to pass through the ammeter – only an amount proportional to it. Note that the polarity light will remain live despite the RCBO being open, so before opening the panel board make sure that the AC shore power connector is disconnected.

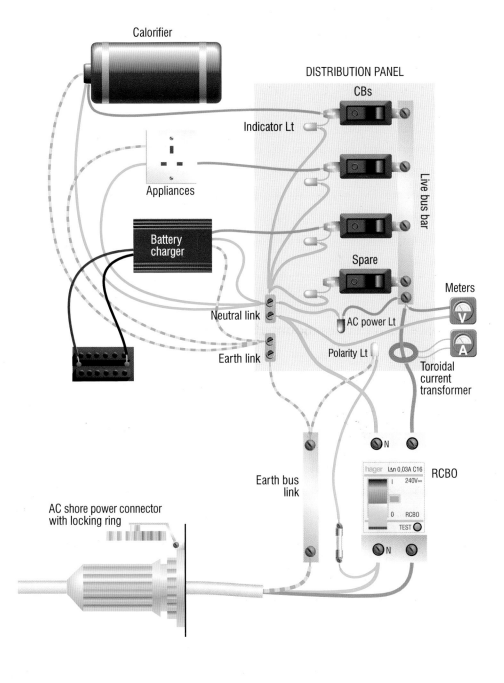

The Full Works

Isolation Transformer

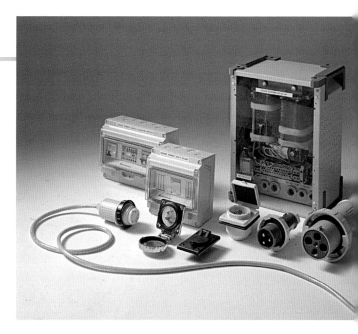

This is a 1:1 transformer which neither raises nor lowers the voltage. Its purpose is to pass the 240V supply through to the boat without there being a direct connection to the shore, the transfer being accomplished by induction between the transformer's coils. This isolation of the input from the output brings two advantages. Firstly, it effectively brings the source of power to within the boat so we can now bring the boat's earth to the neutral point of the transformer. Any earth fault current in the boat's system now only has to travel to this point, making the system both safer and more efficient. Secondly, because the boat's earth is shielded and isolated from the shore earth network, the shore earth isn't carried on to the boat's distribution system and the boat is therefore isolated from leakage or stray currents.

Unfortunately, isolation transformers are heavy (about 30kg) and expensive, which probably explains why only about 5% of offshore craft are fitted with them.

Above: a variety of waterproof shore power plugs overlooked by the heavy bulk of an isolation transformer. To the left of this is the soft start which is designed to prevent the system breakers from tripping as the transformer absorbs the huge surge of current upon connection.

Soft Start

The induction coils of an isolation transformer are huge, and to electrically initiate the flux to oscillate at 50Hz demands an enormous amount of current, which would certainly trip the over-current element of the RCBOs. In mechanical terms, this would be like trying to move off in a car in fourth gear.

What's needed, therefore, is an electrical gear box, and this is where the soft start switches in. Essentially, this is a large resistor straddling a switch on the live cable. When the AC current is switched on, the resistor absorbs most of the initial current until the time delayed relay cuts in, closing the soft start switch - thus shorting out the resistor - to provide the full service current.

Circuit Breakers

Isolation transformers determine the location of protection devices. To detect earth leakage currents, RCBOs and RCDs must be downstream of the transformer. The location of MCBs relative to the isolation transformers can be an issue of discussion, and as such the MCBs will either be seen upstream or downstream of the transformer. A majority view seems to prefer MCBs downstream of the transformer since there is some concern that the transformer's impedance may limit any fault current to the MCB.

See Appendix F for Impedance

Inverters

If you are the kind of owner that would aspire to a limited amount of AC power on the boat, without the hassles of shore power cabling and a distribution system, then consider the inverter.

An inverter is a device which converts a 12V DC supply into 240V AC, thus allowing your battery to mimic (at fairly low amps) your home supply. The inverter passes your 12V DC through a simple oscillating circuit which alternates it between +12V to -12V at 50Hz. This square wave current then goes on to a step-up transformer which raises it to 240V AC, before it is cleaned and filtered to provide a more sinusoidal supply.

You can buy inverters as stand-alone units or combined with a battery charger. They can work in both directions (240V AC to 12V DC and vice versa) but not in opposite directions at the same time. In practice this means that in harbour the unit can feed the batteries and supply the 12V float charge required for domestic needs and, at sea, the role could be reversed, with 240V being delivered to power, say, a microwave oven, television set or computer.

If the inverter is going to be incorporated into a fixed AC shore power system, then it is absolutely vital that the inverter AC output and the shore power AC should never meet - if they did you'd be the sponsor of a very spectacular firework display! This calls for an interlocking changeover switch, usually mounted on the AC panel board. In some cases this may be a fully automatic function, providing a plug-in-and- forget feature.

Since an inverter is a source of AC power, then like the isolation transformer, its neutral is earthed and fed to a common earth link.

Distribution Panel

Compared to its DC counterpart, the AC panel board is very similar in its organisation and layout. The CBs are themed and grouped and the usual system voltage and current is displayed. A very important and additional display feature is the polarity indication light. This light indicates if the polarity of the supply is correct or incorrect (see Poles Apart P.104). The warning symbols on the panel (see diagram in Appendix D) are not there for graphical aesthetics; they are there by the laws of the Recreational Craft Directive.

The location of the MCB/RCBO is vital if it is located off the panel, for either one acts as the main breaker to the boat. It must be remembered that it is not enough to open the breaker before opening up the panel board - for the polarity light will still be live - a complete disconnection from the shore power at the pedestal is required.

The interlocking changeover switch not only separates the inverter AC and shore power AC, but also distributes the correct amount of power to the AC distribution system accordingly. Inverters aren't capable of driving such power-hungry devices as immersion heaters and conventional ovens. Once the inverter is selected, the only circuits capable of being sustained by the inverter are on line. With shore power selected, all circuits and appliances are on line. For this reason, the CBs will be grouped into two sections.

For a digital display of voltage and current, as in this example, the AC information is rectified by the nest of diodes before being passed to the digital meter. A toroidal current transformer may be used to supply the signals for a current reading.

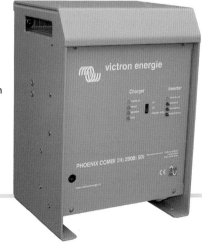

Top: the simplest inverter - 12V from the battery enters from the red and black cables, and a single output socket provides 240V AC current.

Left: the inverter/charger combination providing two functions in one box. However they can only do one or the other, not both at the same time.

Poles Apart

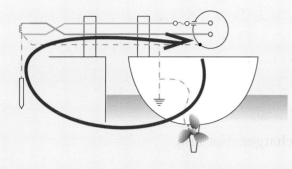

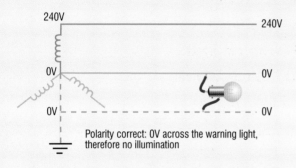

Polarity correct: 0V across the warning light, therefore no illumination

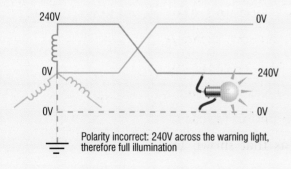

Polarity incorrect: 240V across the warning light, therefore full illumination

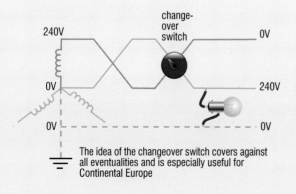

The idea of the changeover switch covers against all eventualities and is especially useful for Continental Europe

The polarity light is one of the most important items on the distribution panel. It tells the owner that the boat's wiring conforms with the shore power, ie live to live, neutral to neutral and earth to earth. Although the light itself is on the panel, its connections to the system must precede everything else to ensure a correct and true indication.

WARNING! – Beware when opening a distribution panel from a bulkhead because the cables feeding the polarity light could be live despite the main CBs being open. To be absolutely safe, remove the shore power connection plug from the marina pedestal as well as opening the main CBs before opening up an AC board panel. Note that the polarity light cables are fused at the source. When you connect your boat to the shore, you should always check that the polarity light shows an 'all clear' before throwing the boat's MCB to the distribution board. If the polarity were incorrect then the neutral would become live and the live neutral (top). If a fault were then to occur, the fuse would be on the neutral side and effectively by-passed (and therefore useless) and the earth would carry the fault current back to the source and along to the fault again – a very hazardous event.

It is essential for every owner to determine first whether an illuminated light means that the polarity is correct or incorrect. The latter seems more logical – a warning that all is not as it should be. Some panels have a pair of lights, ie polarity correct and polarity incorrect, and you have to deliberately look at which one is illuminated to check the polarity condition, which seems to leave room for possible confusion.

The occurrence of an incorrect polarity in UK marinas is highly unlikely. Once on the Continent, however, the situation is very different and could be quite a regular occurrence. To rectify this, some experienced owners install a changeover switch almost immediately after the shore power receptor, so that should a faulty polarity condition occur, then the owner simply switches over to change the polarity in accordance with the boat.

If you have an isolation transformer fitted, then there's no need for any warning provision as a change in shore power polarity will have no effect upon the transformer's output.

Practical Considerations

Let's face it, in the real world there are very few boats which have such things as isolation transformers, galvanic isolators and dedicated earth ground plates, or indeed have their AC appliances and all underwater metalwork earthed and bonded to ground potential. Most boats simply use a flexible extension cable connected principally to their battery charger, but also perhaps powering a fan heater or dehumidifier.

With this in mind, a set-up similar to that shown on P. 98 would suffice. Here, reliance is placed on an RCD as the primary protection device, with some additional safeguards from having the earth cable carrying any fault currents back to the source ashore. Indeed, the British Marine Electronics Association (BMEA) is just short of committing themselves to incorporating this recommendation into their Code of Practice.

Similarly, the debate about bonding and grounding AC appliances and underwater fittings still rumbles on. Although theoretically correct, it is generally agreed that bonding and grounding leads to other problems. However, an electrician I consulted put it to me that, if I was the owner of a boat with an extensive AC system, would I be happy touching a stainless steel sink full of water, knowing that it wasn't bonded to earth potential? If not, then an installation such as that shown in Appendix E would seem to make sense.

Websites

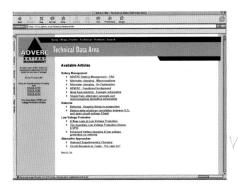

www.adverc.co.uk

www.balmar.net

Adverc's main mark is their intelligent charge controller that enhances the performance of an alternator, without stressing the alternator or battery. Adverc's broad and specialist field is battery management, covering all aspects of battery charging and monitoring.

Their website has its own novel search engine for finding particular items of interest and the site is very educationally biased, attempting to provide knowledge and options concerning on board battery related matters.

It contains easy to understand technical articles with Q & A pages, including ideal boat schematics.

Balmar is an American outfit which is becoming a sizeable force and influence in the EU. UK dealers stock mainly their alternators and superb smart regulators, but once you tour their website, it's plain to see their large and diverse range of products. Their site offers Q&A, troubleshooting FAQs and the availability to download all their product manuals. Balmar's technical team also offers an online technical support service which operates during the working hours of Pacific Standard Time. The site also hot links you to battery and engine manufacturers, and marine electrical gear and retailers amongst many others.

www.ampair.com

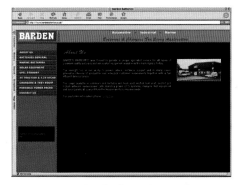

www.bardenbatteries.co.uk

If you need alternative and natural sources of power for your boat, then **Ampair** has everything you need. Famed and renowned for their wind generators, they also offer solar and water generator alternatives backed up with all the required voltage regulation and monitoring.

The website allows browsers to see many examples of applications and provides a comprehensive source of technical information as well as detailed product and installation specifications. Ampair's catalogue is available to download, and the site also offers links to other sites in their field.

Barden Batteries are marine battery and solar specialists and provide a wide range of product and technical support throughout the UK and EU. Their marine battery folio includes: Red Flash, Optima, Sonnenschein and Deta gel, Lifeline AGM and Delphi Marine, Yuasa sealed, Deta, Hankook and Baron wet lead-acid. Barden are also the sole UK distributors for the German manufactured SOLARA solar panels, which are available in both conventional glass and the unique semi-flexible marinised panels, which are self cleaning and can be walked upon.

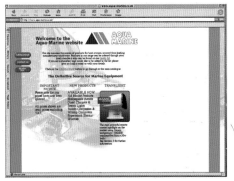

www.aqua-marine.co.uk

www.dg2k.co.uk

AquaMarine is a major distributor of electrical equipment and other chandlery from around the world. The equipment available is based on well made quality items.

The website can display their full catalogue, and section 3 of the catalogue covers all the electrical equipment. Further information for each electrical item can be obtained, which often hot links you straight to the manufacturer's site to provide you with first hand information and data.

This is **Driftgate 2000 Ltd**, who concentrate on battery charging and charging systems. Their champion product is the 'X-split'. This is a split charging device which electronically emulates a diode splitter, the benefit being no voltage drop across the charge splitter - less than 100mV compared with 0.7V across a diode.

There is nothing to download from their website, but their complete product range is featured with all the necessary technical data.

www.bluesea.com

Blue Sea Systems design and manufacture electrical circuit components for AC and DC systems aboard boats. Their products for circuit protection, connection, switching, metering and insulating meet UL, CE, SAE and ABYC standards and are available through distributors worldwide.
The website offers customers access to the company's technical expertise in marine electrical systems and includes online and downloadable PDF files for information including catalogues, circuit diagrams, instruction manuals, electrical standards and technical data relating to marine electrical systems.

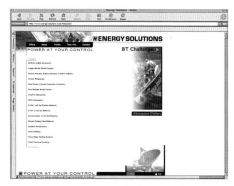

www.energy-solutions.co.uk

Energy Solutions specialise in electrical power for the marine industry and offer specialist advice and services to assist in bringing their products together into a successful and reliable package. Their website has an impressive range of products and information. Each product is backed up with technical files and brochures, all of which can be downloaded as a PDF file. As well as these there are more general technical files which cover matters like worldwide electrical power standards, charge state tables, wiring standards for the Recreational Craft Directive, voltage drop calculations and metric cable size to AWG conversion tables.

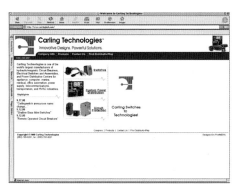

www.carlingtech.com

Carling Technologies is in direct competition with ETA making it very difficult to compare the two products. Needless to say, both are just as reliable as the other in quality and reliability. Carling's website is not so forthcoming with the information as ETA, but they make up for it with the ability to download their CB catalogue and some good PDF download files on some custom solutions which offer sound advice to any boat owner wanting to know about protecting their boat's wiring.

www.etacbe.com

If you already have or would like to aspire to having CBs on your boat, then there's a strong chance it will either be an **ETA** or a Carling circuit breaker. These two firms operate neck and neck and are fiercely competitive, making it impossible to place one against the other. ETA's marine sector covers only a fraction of their overall business, but that isn't to say that most of their product range is irrelevant to boat owners. Both thermal and magnetic CBs are available but RCD/RCBOs are not available.
All the CBs can be browsed on one big page which is very useful and user friendly, and a data sheet PDF file on each individual CB can be downloaded.

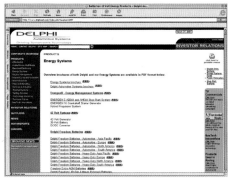

www.delphiauto.com

Delphi is a large American technical engineering company that happens to make very good marine batteries which are available through selective outlets in the EU and the UK. Once you enter their website, go to 'Energy Systems' under their products menu to reveal their 'Freedom' range of batteries. Within the Freedom range are the marine batteries which can be viewed by downloading or browsing the PDF files. The information available is very comprehensive as well as educational, and as such offers probably the best site as far as batteries are concerned.

www.heartinterface.com

Heart Interface is an American firm whose products appear in the UK through selective outlets. Better known for their 'Link' battery monitor, they also make the 'Pathmaker' relay - both of which are featured in this book. Their website can take you to their marine section where you're faced with a comprehensive range of 'Freedom' chargers, inverters and combi-units with all their accessories. Next to that is a very worthwhile list of FAQs which will answer all your battery and charging system needs.
As with most American websites, all the manuals of every product can be downloaded as a PDF file.

Websites

www.kurandamarine.co.uk

Kuranda Marine distribute a full range of electrical products, heating and cooking systems for inland and offshore marine applications. They can offer complete system solutions, from the smallest inverter to an automatically managed generator, shore power and parallel inverter system with split charging and full battery monitoring.

The site offers an overview of all the products and services available from Kuranda Marine along with contact details for on or offline enquiries.

www.mastervolt.com

Mastervolt is a Dutch company with a UK and worldwide agency network. Very highly regarded, it perhaps leads the way in offshore and inland AC and DC distribution systems. The site offers an on-line illustrated catalogue with facilities to order goods, or you can download their complete catalogue. There are also numerous information and technical PDF files available to download as well as some Q&A pages of problems familiar to many boat owners.

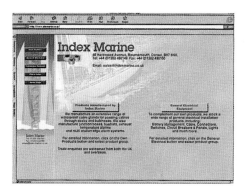

www.indexmarine.co.uk

Index Marine is a highly respected UK manufacturing and distribution company specialising in marine electrical equipment. Their own manufactured products include the successful range of Index Marine cable glands, junction boxes and busbars. The website covers the complete range of marine electrical products including tinned and standard cable, terminals, waterproof connectors, switches, circuit breakers, panels and much more. The site gives full specifications, illustrations and prices. Index Marine are also able to source specialist products to meet individual requirements.

www.the-merlin-group.com

Merlin are famed for their popular display board which they take to all their boat shows. The large display board mimics a yacht's electrical system and is fully interactive to help the public understand the workings of an electrical system. The Merlin site provides a basic preview of the products on offer, otherwise you can be hot linked to their main site at www.power-store.com where you can shop for the goods on line. Each of their product categories is backed up with FAQs to put your mind at rest before ordering.

www.marlec.co.uk

Like Ampair, their main competitor, Marlec specialise in alternative sources of power and brand their products as 'renewable energy'. Marlec's own range of Rutland Windchargers have been used in leisure and professional marine applications since 1979 and have proved themselves ever since. Solar panels by Solarex, also form part of their product range. The website takes you through their complete product folio with specifications, system diagrams and brochures to download.

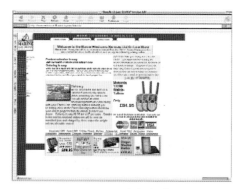

www.mesltd.co.uk

Marine Electronic Services is much more of an electronic online shopping service than anything else, however they do have some electrical products like instruments, chargers and inverters.

For those who are lost and confused when it comes to electronic aids and gadgets, then this website has some '10 minute guides' on radar, GPS, plotters and PC charting. These short guides, which are designed to be read in 10 minutes, should put you in the know, before you start shopping around.

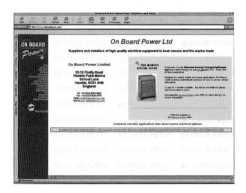

www.onboardpower.com

www.vetus.nl

On Board Power Ltd offer a complete installation and design service to private boat owners and the trade. Their service includes all phases of the project, from initial evaluation of the customer's requirements, through to the commissioning of the final installation.

Their website covers a wide range of products, and in most cases you're hot linked straight to the manufacturer's site. Take a look at the 'data downloads' on the site and also browse the 'wire' sub-section where you can find some very useful tables and charts on wiring.

Vetus can supply quality marine electrical equipment, Including: batteries, chargers, inverters, diesel generators, diode splitters, management systems, high output alternators, switches, cables, panels, interior lights, instruments and gauges. They also supply many other different products from diesel engines to fenders.

The multi-language site contains the entire illustrated product catalogue. There are also installation instructions available to download, to help you choose the right equipment.

www.sowester.com

www.victronenergie.nl

This company is now **Sowester Simpson-Lawrence**, which makes it a formidable force in electrical yacht chandlery. Their website is awesome when it comes to the range and diversity of products, not to mention some very strong brands like ETA, Carling, LVM, Blue Sea, Delco, Webasto and Eberspacher. General outline information comes with each product but disappointingly you're not able to be hot linked to the manufacturer where more information could be obtained.

Like Mastervolt, **Victron** is another highly regarded Dutch company, hence the two of them compete for the same territory. Although Mastervolt cover a wider field, Victron specialise in chargers, inverters and ultra quiet generators which are proving, pound for pound, to be cheaper than shore power. The website has a complete product listing where one can download the date sheets and product manuals. Technical information is also available on the site as browser articles or PDF download files.

www.taplininternational.co.uk

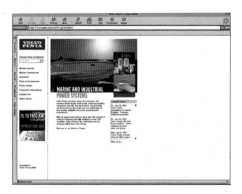

www.penta.volvo.se

Taplin have been manufacturing marine electrical panels for 100 years. Their modern factory is equipped with the latest computer controlled profiling and engraving machinery. They supply production engine instrument panels and AC/DC distribution systems for major boat builders in the UK and overseas. Police, harbour, customs vessels and MOD special craft come within their expertise. Taplin also specialise in the manufacture of bespoke panels for individual yachts and motor craft in a variety of types and finishes.

Volvo Penta is the only engine manufacturer with a website which has recognised that there is an electrical system to an engine. In the 'parts and accessories' section of their site you'll find 'electrical' and 'instrumentation' headings. These are basically PDF download files of their catalogue which is always freely available at the boat shows. However, there is good information there and chapter 3 of this book has adopted much of the Volvo principles.

Bibliography

• *The Marine Electrics Book*
by Geoffrey O'Connell,
published by Ashford, Buchan & Enright

• *Boat Electrical Systems* by Dag Pike,
published by Adlard Coles Nautical

• *Boat Electrics* by John Watney,
published by David and Charles

• *Engine Monitoring on Yachts* by Hans Donat,
published by VDO marine GmbH

• *Simple Boat Electrics* by John Myatt,
published by Fernhurst

• *Reliable Marine Electrics* by Chris Laming,
published by Adlard Coles Nautical

• *The Marine Electrical & Electronics Bible*
by John Payne, published by Adlard Coles
Nautical

• *12V Bible for Boats* by Miner Botherton,
published by Waterline

• *Boatowner's Mechanical & Electrical Manual*
by Nigel Calder, published by Adlard Coles
Nautical

• *12V Doctor's Handbook* and *12V Doctor's
Trouble Shooting Book* by Edgar Beyn,
published by C. Plath

• *Boatowners Wiring Manual* by Charles Wing,
published by Adlard Coles Nautical

The last four books are from the United
States and, although arguably the best books
on the list, some of the practices and
procedures differ from those in the UK and
in EU countries - especially where AC shore
power is concerned.

Index

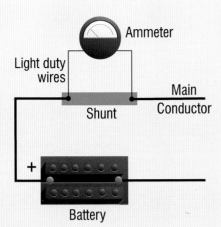

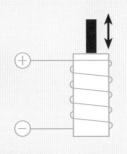

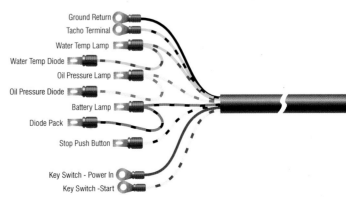

APPENDIX C

Heavy switching circuit

Light signal circuit

SHUNT

The shunt has a low resistance of a known value such that a tiny voltage is dropped across it. As the voltage drop is proportional to the current flowing through the shunt, the millivolt value sensed across the shunt can be calibrated to represent current. This idea reduces the amount of main conductor wiring, should the ammeter be located remotely. If the meter is remote from the shunt, then the long light duty wires will not cause any nuisance radio interference.

SOLENOID

This is a wire coil wound around an insulated cylinder, inside of which is a solid steel rod, which can move freely within it. An application of a current to the coil will, through the electromagnetic influence, cause the steel rod to be drawn into the cylinder coil in a linear fashion. The inward pull on the rod is independent upon the directional flow of current in the coil. Indeed, exactly the same inward pull upon the rod would occur if an AC current was applied to the coil.

RELAY

A relay is an extension of the solenoid, in that the linear movement of the steel rod activates a switch either to open or close a circuit completely independent and separate from the solenoid coil.

BI-METALLIC SWITCH

This switch incorporates a strip of dissimilar metals bonded together. With changes in temperature the different rates of expansion cause the strip to bend, either opening or closing a circuit, and acting as an 'on-off' switch.

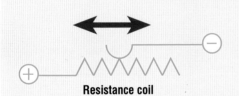

Resistance coil

RHEOSTAT/VARIABLE RESISTOR

If this device is coupled to an electrical circuit we can incorporate a gauge to pick up any changes in the current as the rheostat alters its electrical resistance within the circuit.

THERMISTOR PELLET

This is a metal whose electrical resistance changes with variations in temperature. If this device is connected to an electrical circuit, we can incorporate a gauge to pick up any changes in the current flow as the thermistor alters its electrical resistance.

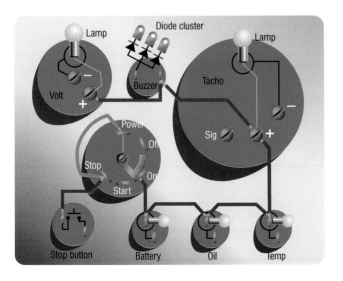

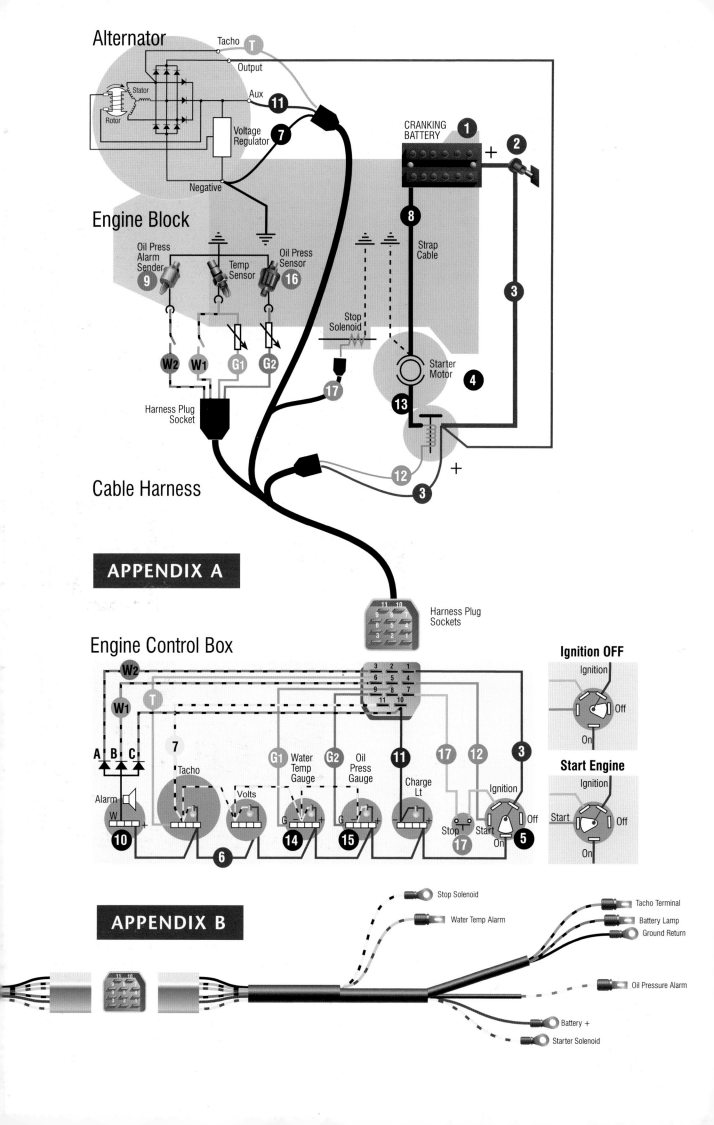

Alternator

Tacho — T

Output

Stator

Rotor

Aux — 11

Voltage Regulator — 7

Negative

Engine Block

Oil Press Alarm Sender — 9

Temp Sensor

Oil Press Sensor — 16

W2   W1   G1   G2

Harness Plug Socket

Stop Solenoid

17

Cable Harness

CRANKING BATTERY — 1

2

8

Strap Cable

3

Starter Motor

4

13

12

3

Harness Plug Sockets

APPENDIX A

Engine Control Box

W2

W1

T

A B C

7

Tacho

Alarm

W — 10

G1 Water Temp Gauge

G2 Oil Press Gauge

Volts

11 Charge Lt

17   12   3

14   15

Stop — 17   Start   Ignition — 5   Off   On

6

Ignition OFF

Ignition

Off

On

Start Engine

Ignition

Start   Off

On

APPENDIX B

Stop Solenoid

Water Temp Alarm

Tacho Terminal

Battery Lamp

Ground Return

Oil Pressure Alarm

Battery +

Starter Solenoid

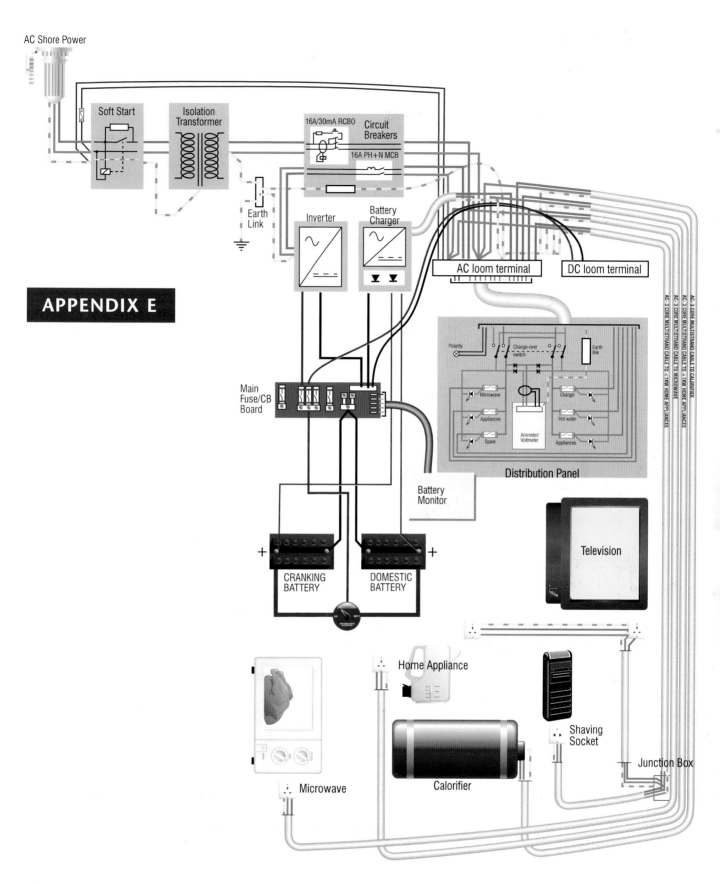

AC Shore Power

Soft Start

Isolation Transformer

16A/30mA RCBO Circuit Breakers

16A PH+N MCB

Earth Link

Inverter

Battery Charger

AC loom terminal

DC loom terminal

APPENDIX E

Polarity

Change-over switch

Earth link

Microwave

Charger

Appliances

Hot water

Spare

Ammeter/ Voltmeter

Appliances

Distribution Panel

Main Fuse/CB Board

Battery Monitor

Television

+ CRANKING BATTERY

DOMESTIC BATTERY +

AC 3 CORE MULTISTRAND CABLE TO <1KW HOME APPLIANCES

AC 3 CORE MULTISTRAND CABLE TO +1KW HOME APPLIANCES

AC 3 CORE MULTISTRAND CABLE TO CALORIFIER

AC 3 CORE MULTISTRAND CABLE TO MICROWAVE

Home Appliance

Shaving Socket

Junction Box

Microwave

Calorifier

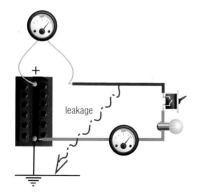

+

leakage

LEAKAGE OR STRAY CURRENTS

Suppose we break the connection at the +ve terminal of a 12V battery as shown, and place a voltmeter across the break. If the switch to the lamp is closed, the lamp will remain extinguished (since the high resistance of the voltmeter absorbs the full voltage drop) and the voltmeter will register 12V and a very slight current will register at the ammeter. With the lamp switch open, we would expect to see a null reading on both the voltmeter and the ammeter. However, if a reading is apparent at the voltmeter, then current is leaking or straying from the battery source. If it so happened that the meter at the battery was a multimeter, we could set it to register current and so establish the size of this current leakage.

DIODE

An electronic non-return valve which ensures current flow in one direction (forward) but not in the other (inverse). Virtually all diodes these days are solid state electronic devices doped with semiconductor materials.

Symbol-

A typical alternator diode would have 50A forward current rating, a 50V peak inverse voltage rating, and 0.6V forward voltage drop

ZENER DIODE

Like a normal diode, this is designed to break down and conduct in the reverse direction at a pre-set value. It is mainly used to stabilise fluctuating voltages.

INDUCTANCE

If we applied a current to a coil it would produce a magnetic flux proportional to the rate of current applied. If the circuit to the coil were broken through opening the switch, then the collapsing current would create a magnetic flux in the coil so as to induce a current in opposition to the collapsing current. Under certain conditions the voltages created from an induced current can be greater than the original voltage applied, such that a spark could arc across the switch.

IMPEDANCE

Placing an induction coil in a DC circuit has little or no effect upon the overall resistance of the circuit. However, once an AC current is passed through the coil, it can pose a considerable resistance in the circuit. Very simply, as the +ve waveform of the AC cycle tends towards zero in the coil, this collapsing current induces a current in the coil to oppose the collapsing supply current, thereby offering resistance to the AC current (see Inductance). This phenomenon is similarly reflected in the coil as the -ve waveform tends towards zero; this sort of resistance is called inductive reactance, sometimes referred to as AC resistance.

This reactance coupled with any resistance that may exist in the coil is jointly expressed as Impedance.

TRANSISTOR

This is an electronic 2-way switch with no moving parts. There are 2 types, NPN and PNP, the former being more common.

An NPN transistor is really two diodes back to back. The current and voltage at the base (B) controls the current flowing from the collector (C) to the emitter (E). A low voltage/current from B to E means virtually no voltage/current from C to E. However a large voltage/current from B to E will cause a current to avalanche from C to E. From here we can see the beginnings of a voltage sensitive switching device or perhaps more relevantly, a voltage regulator.

Symbol for an NPN transistor

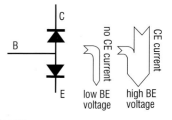

APPENDIX F

ANODE/CATHODE

These are often the cause of great confusion, since school day physics has embedded in our minds that the anode is positive and the cathode is negative. Strictly speaking this isn't true for the nomination of an anode or a cathode is dependent upon the direction of the current and not the polarity of the voltage. By observing the Galvanic Series Table below we can graphically illustrate this phenomenon. As the table shows, zinc is virtually the most anodic material in the series, yet we also see that it is virtually at the negative end of the voltage spectrum. This means that if zinc is paired off with any other of the dissimilar metals it will take on the negative pole relative to the other material and yet still be labelled as an anode. Since conventional current flow is from positive to negative, we can conclude that the anode will always receive the current from the external circuit. This phenomenon stands true for zinc-carbon torch batteries, electrolysis (familiar with electro-plating metals) and vacuum tube and silicon diodes. Interestingly for a rechargeable battery, for example, as it cycles from charging to discharging, the polarity of its terminals remains the same but the anodes and cathodes charge over.

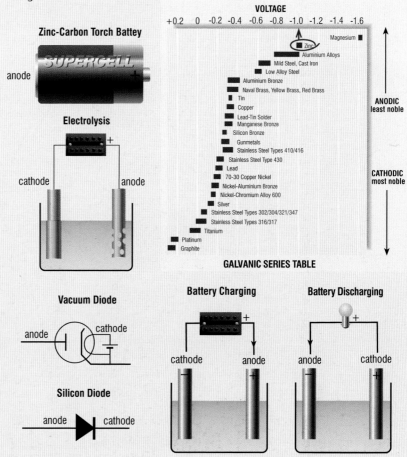